CLINICAL JUDGMENT: A CRITICAL APPRAISAL

PHILOSOPHY AND MEDICINE

Editors:

H. TRISTRAM ENGELHARDT, JR.

Kennedy Institute of Ethics, Georgetown University, Washington, D.C., U.S.A.

STUART F. SPICKER

University of Connecticut Health Center, Farmington, Conn., U.S.A.

VOLUME 6

CLINICAL JUDGMENT: A CRITICAL APPRAISAL

PROCEEDINGS OF THE FIFTH TRANS-DISCIPLINARY SYMPOSIUM ON PHILOSOPHY AND MEDICINE HELD AT LOS ANGELES, CALIFORNIA, APRIL 14–16, 1977

Edited by

H. TRISTRAM ENGELHARDT, JR.

Kennedy Institute of Ethics, Georgetown University, Washington, D.C., U.S.A.

STUART F. SPICKER

University of Connecticut Health Center, Farmington, Conn., U.S.A.

BERNARD TOWERS

University of California at Los Angeles, California, U.S.A.

D. REIDEL PUBLISHING COMPANY

DORDRECHT : HOLLAND / BOSTON : U.S.A.

LONDON : ENGLAND

Library of Congress Cataloging in Publication Data

Trans-disciplinary Symposium on Philosophy and Medicine,
5th, Los Angeles, 1977.
Clinical judgment.

(Philosophy and medicine ; v. 6)
Includes bibliographies and index.
1. Medical logic—Congresses. 2. Medicine, Clinical—Decision making—Congresses. I. Engelhardt, Hugo Tristram, 1941– . II. Spicker, Stuart F., 1937– III. Towers, Bernard. IV. Title.
R723.T7 1977 616'.001'9 78–26504
ISBN 90–277–0952–1

Published by D. Reidel Publishing Company,
P.O. Box 17, Dordrecht, Holland

Sold and distributed in the U.S.A., Canada, and Mexico
by D. Reidel Publishing Company, Inc.
Lincoln Building, 160 Old Derby Street, Hingham,
Mass. 02043, U.S.A.

Printed in The Netherlands

EDITORIAL PREFACE

Over a period of a year, the symposium on clinical judgment has taken shape as a volume devoted to the analysis of how knowledge claims are framed in medicine and how choices of treatment are made. We hope it will afford the reader, whether layman, physician or philosopher, a useful perspective on the process of knowing what occurs in medicine; and that the results of the discussions at the Fifth Symposium on Philosophy and Medicine will lead to a better understanding of how philosophy and medicine can usefully challenge each other. As the interchange between physicians, philosophers, nurses and psychologists recorded in the major papers, the commentaries and the round table discussion shows, these issues are truly interdisciplinary. In particular, they have shown that members of the health care professions have much to learn about themselves from philosophers as well as much of interest to engage philosophers. By making the structure of medical reasoning more apparent to its users, philosophers can show health care practitioners how better to master clinical judgment and how better to focus it towards the goods and values medicine wishes to pursue. Becoming clearer about the process of knowing can in short teach us how to know better and how to learn more efficiently. The result can be more than (though it surely would be enough!) a powerful intellectual insight into a major cultural endeavor, medicine. The result can be better diagnosis and treatment of patients and more successful education of health care professions. Being clearer about ideas means also being clearer about the actions we take to be of great significance and which are directed by such ideas. And in this process not only is medicine enriched, but philosophy is as well, by an investigation of substance and social importance – not to mention of conceptual interest.

Though the term medicine is used throughout this volume, it is meant broadly – somewhat as philosophy is used in the degree Doctor of Philosophy. One should see it encompassing what is embraced by the more cumbersome phrase, "The health care sciences, arts, and technologies." It includes then not only medicine *in sensu stricto*, but such endeavors as nursing, occupational therapy, dentistry, veterinary medicine, clinical psychology, psychoanalysis, and the practice of physician assistants. All of these enterprises, as well as others, are concerned with categorizing the complaints of patients (clients) in order to forward accurate prognoses and enter upon useful courses of treatment. Moreover, they all share in the fact-value ambiguities of health and disease language. In describing individuals as ill or well, healthy or diseased, they also evaluate them. Moreover, in terming individuals sick, health care professionals cast patients (clients) into social roles with vast implications – consider for example Talcott Parson's sick role [1]. Clinical judgments are not only intellectual exercises with the usual practical consequences we envisage (e.g., good or bad treatment). Clinical judgments also have consequences that impinge on general social goods; they respond as well to our evaluations and our social structures. As in the other volumes in this series on philosophy and medicine, we see evaluations intertwined in medical explanations. The result is that no adequate sense can be made of bioethical problems without understanding the particular enterprises of knowing which constitute medicine.

The essays in this volume are offered as one step in understanding the theory of knowledge in medicine. They are meant as a contribution to a philosophy of medicine that seeks to engage some of the more ambitious themes of the first volume in this series – *Evaluation and Explanation in the Biomedical Sciences*. Much more needs to be done, and will be done in this series, in the way of exploring such subjects as the role of evaluation in medical explanations, the value-laden character of the language of disease and health, and the hidden imperatives of the sick role. Much more must be undertaken with regard to understanding the ambiguities of disease language. This is particularly the case in coming to terms with nosography and the generation of nosologies in the use of computers in medicine. Moreover, the problem of the logic and the psychology of discovery in medicine remains almost untouched.

This volume offers a first approach towards the analysis of the nature of clinical judgment. We are grateful for the support of the School of Medicine of the University of California at Los Angeles, the Research Foundation of the American Medical Association, the Franklin J. Matchette Foundation, and D. Reidel Publishing Company, who made this beginning possible. Not only are we grateful for their financial support, but for their moral support as well, which enabled us to hold the symposium which was the ancestor of this volume. Moreover, we stand in special debt to the School of Medicine of the University of California at Los Angeles for the many ways in which its administration encouraged and made possible this meeting of clinicians and scholars interested in understanding clinical judgment. Many individuals labored unselfishly in the preparation and conduct of this symposium. We are deeply grateful to them all. Here we can name but a few: Sherman M. Mellinkoff, Robert Adams, Rodney Bluestone, Susan Engelhardt, Sharon Hill, Marjorie Huffman, Dwaine Lawrence, Norman Levan, Charles Lewis, Ruth Walker Moskop, Fritz Redlich, Arthur Rivin, Bernice Wenzel, Paula Wilkes, and William J. Winslade. A special thanks in this regard is due to Jane Backlund, who inherited the post-symposium red tape that comes in the wake of any large conference. We are also in debt to her for the careful preparation of the manuscripts of this volume and for the generation of many of the references. Finally, we must acknowledge our debt to the many individuals who attended this symposium on clinical judgment and helped through their discussion to frame the papers in the final form in which they appear in this volume.

April 21, 1978

H. TRISTRAM ENGELHARDT, JR.
STUART F. SPICKER
BERNARD TOWERS

BIBLIOGRAPHY

1. Parsons, T.: 1967, *Sociological Theory and Modern Society*, Free Press, New York.

TABLE OF CONTENTS

SECTION IV / JUDGMENT AND METHODS IN CLINICAL JUDGMENT

H. TRISTRAM ENGELHARDT, JR.

INTRODUCTION

> Clinicus is also used for a physician – In regard [to] physicians [who] are much conversant about the beds of the sick. Clinic is now seldom used but for a quack; or for an empirical nurse, who pretends to have learned the art of curing diseases by attending on the sick.
>
> Ephraim Chambers, *Cyclopaedia; or an Universal Dictionary of Arts and Sciences* (1751)

As the definition of clinician in Chambers' dictionary shows, being called a clinician does not appear to have always been an unalloyed compliment. We, though, take it to be unequivocally a term of honor, designating a physician skilled in the treatment of patients; consider the force of saying "he is no clinician". Chambers' skepticism concerning clinicians was due as the entry suggests to a suspicion of empirics, physicians who practiced without an adequate theoretical basis for their medical judgments. In part, current interest in and criticism of clinical judgment stems from a similar distrust of the empiric. There is a suspicion that the judgments of clinicians would be more reliable if rendered more rational by following explicit rules and recipes for diagnosis and treatment. There is, as well, interest in clinical judgment in order to understand how to reconstruct its successes and have them matched by undertakings of artificial intelligence – computers. Yet others would see in clinical judgment, decisions made on the basis of an intuition born of years of bedside experience, and not reducible to explicit formulae. The interest in clinical judgment is thus diverse if not contradictory: there is interest among clinicians and others in reducing the basis of medical decisions to formal rules as well as interest in denying that such rules would match the success achieved by the intuitive decisions of clinicians.

The phrase 'clinical judgment' is in other ways ambiguous. On the one hand, it may be taken to refer to a capacity to make medical judgments concerning patients. One might think here of the famous line from *Epidemics*, "Declare the past, diagnose the present, foretell the future; practice these acts. As to diseases, make a habit of two things – to help, or at least to do no harm." (*Epidemics* I, 11 [11].) Clinical judgment in this very traditional sense

H. T. Engelhardt, Jr., S. F. Spicker and B. Towers (eds.), Clinical Judgment: A Critical Appraisal, xi–xxiv. *All Rights Reserved.*

is the ability to form diagnoses, forward prognoses, and make choices of treatment which help the patient or which at least do him or her no harm. Here the judgment is 'clinical' in being focused on patient problems. Clinical judgment, on the other hand, is often used to identify the experiential origins of such a capacity of judgment – bedside experience, learning at the κλινη. In this sense clinical judgment is often used as a shibboleth of physicians, or as a final justification of a diagnosis or choice of treatment. One finds physicians asserting that one can make adequate clinical judgments only on the basis of actual experience, not simply on the basis of general principles of physiology, pharmacology, pathology, or even on the basis of a reconstruction of past clinical judgments. The claim is often that the knowledge gained on the basis of years of clinical experience is not reducible to explicit rules, recipes, or basic principles.

One may view reliance on actual bedside experience as opposed to recipes for diagnoses as a remnant of the Hippocratic rejection of metaphysical speculation and the embrace instead of an empirical understanding of medicine – one should learn medicine by observing patients. Such opinions are expressed by clinicians like Thomas Sydenham (1624–1689): "In writing the history of diseases, every philosophical hypothesis whatever, that has previously occupied the mind of the author, should be in abeyance." ([27], p. 2). Yet one must recognize that such rejections of metaphysical and philosophical speculation have rarely been taken to include a rejection of rational or logical principles in medical decision-making. The author of the *Precepts*, for example, enjoins physicians to use "experience combined with reason" and to engage in theorizing which "lays its foundation in incident and deduces its conclusions in accordance with phenomena." (*Precepts* I, [12].) In fact, there has been a rather traditional concern to understand medical decision-making and render it more rational. One should think here of Sydenham [27], Sauvages [23], or Cullen's [5] concern about how physicians should use their clinical experience, their observation of patients, as a basis for developing reliable diagnoses, prognoses, and successful approaches to treatment.

In fact, beginning with Gilbert Blane's *Medical Logic* [3] the nineteenth century produced a series of increasingly disciplined reflections on medical reasoning. When one examines such works, one sees a concern to outline the rules for successful reasoning in medicine similar to the concern which motivated many of the essays in this volume. Oesterlen, for example, in his *Medical Logic*, sought to apply the logic of John Stuart Mill to medicine. Oesterlen wished "to show, clearly and impressively, the mode in which we must proceed in our observations, investigations, and conclusions, in order that our

Theorems and Problems may become more clearly intelligible, and that we may arrive at experimental truths and definite laws in our department of science, as well as at scientific principles of practice." ([17], p. vii). Oesterlen was seeking a way of rendering more explicit, rational, and empirical the processes of medical observation and clinical judgment. "Thus, when we speak of Medical Logic, we simply mean the application of the science in question to our own special province. We only seek by its aid to explain and develop more clearly the ways and means, the various processes by which the natural philosopher, the physiologist, and the physician are accustomed to proceed in their investigation of phenomena, occurrences, and influences." ([17], p. 4). Similarly, Bieganski, in his *Medical Logic*, sought a disciplined analysis and reconstruction of the processes of medical reasoning, the provision of a scientific medical theory of knowledge. Bieganski wished to place the accent upon the practical task of medical knowledge and to provide a "Theory of medical diagnoses and of the indications for the treatment of patients," ([1], p. 6)[1] much as we have sought in this volume. In fact, Bieganski wished to develop a critique of medical knowledge, as the German translation has it, a "Kritik der medizinischen Erkenntnis," a "Kritik der Methoden ärztlicher Erkenntnis" ([1], p. 14). In such attempts many usually intertwined goals have been pursued: 1) a rational reconstruction of the processes of clinical diagnosis and decision-making, with a view to understanding the nature of securing the validity and truth of such endeavors, 2) a psychological account of the processes of clinical judgment with a view to understanding the nature and securing the efficiency of such judgments, and 3) an epistemologically or psychologically based revision of actual clinical approaches to medical diagnosis and decision-making to the end of greater validity, truth and efficiency in clinical judgment. The third has often been seen to depend upon the first and the second. In order to forward a revision of clinical approaches to diagnosis and decision-making, an account of the nature of clinical judgment has been sought. The extent of past reflections on these issues is impressive, as a mere review of Bieganski's Introduction shows. The tradition of philosophical and medical interest in clinical judgment is extensive.

The rational reconstruction and explication of the processes of clinical judgment have faced difficulties for a number of reasons. As suggested above, a reconstruction of clinical judgment has not always been well received by physicans. Clinicians are often by temperament ill-suited to formal logical and mathematical accounts of medicine. Frequently they do not have a high regard for psychological accounts. There has as well been considerable difficulty in gaining sufficient precision in the language used in talking about clinical judgment (not to mention basic empirical knowledge in medicine) to allow, until recently, much success or utility in such ventures. This has impeded an unambiguous reconstruction of the processes of clinical judgment and has been nowhere more apparent than in attempts to employ computers for clinical diagnosis. The result has been a return to eighteenth century interests in nosography and nosology and to philosophical concerns with the definition of disease. As Proppe remarked – "The problems of a general computer diagnostic [approach to clinical diagnosis] resides in the unclarity of what should in fact be diagnosed. If one really means, by this, diseases differentiable by means of individual symptoms or groups of symptoms or even in contrast to one another, then one must try to say by which criteria one can recognize when an appearance in nature presents a particular disease and when not." ([20], p. 131.) As a consequence of such concerns for greater clarity in the generation of medical data, disciplines for describing patients' complaints have been elaborated such as the problem-oriented medical record ([2], [14]). The literature has also again begun to swell with volumes dedicated to analyzing medical reasoning in the clinical context. One might think here, for example, of Alvan R. Feinstein's *Clinical Judgment* [7], Lee B. Lusted's *Introduction to Medical Decision-Making* [14], Edmond A. Murphy's *The Logic of Medicine* [16], Henrik R. Wulff's *Rational Diagnosis and Treatment* [31], Rudolf Gross's *Zur klinischen Dimension der Medizin* [10], Karl E. Rothschuh's *Prinzipien der Medizin* [21], Wolfgang Wieland's *Diagnose: Überlegungen zur Medizin Theorie* [30], and Hans Westmeyer's *Logik der Diagnostik: Grundlagen einer normativen Diagnostik* [29]. The goal has been an analysis of the "tacit knowing" of medicine. Though such analysis is fraught with dangers, as contributors to this volume will indicate, it also has virtues. As Michael Polanyi put it, "The meticulous dismembering of a text, which can kill its appreciation, can also supply material for a much deeper understanding of it." ([19], p. 19.) Though probably not all tacit knowledge should be rendered explicit, most of the tacit knowledge appealed to by physicians in clinical judgment can probably benefit from an analytic regard.

Many proposals have been forwarded to formalize the language of reflection upon clinical judgment in order to render clinical judgments more accurate, replicable and independent of the idiosyncracies of the particular physician. Consider, for example, Sadegh-zadeh's description of the process of clinical judgment.

Proceeding from a non-empty set of data D at a time-point t_1, which we will call D_{t_1}, the physician chooses from the infinite set A of all possible *actions* which one could consider, the particular sub-set which we will call A_1, and accomplishes it. The results of these actions carry the original set of data D_{t_1} over into the set of data D_{t_2} at the time-point t_2. Proceeding from D_{t_2} a second sub-set A_2 is chosen from A and rendered actual, etc. The finite sequence of this acquisition of data at time-points $t_1 \ldots\ldots t_n$ and the selection of actions

$$D_{t_1}, A_1, D_{t_2}, A_2 \ldots . D_{t_n}, A_n$$

is called the 'clinical process', where n $\geqslant$ 1. The fundamental problem of the clinical process is the question: is it possible to construct an effective procedure p which can be initiated at time-point t_1 concerning the data set D_{t_i} available concerning a patient where $1 \leqslant i \leqslant n$ so that the optimal set of actions A_i can be selected unambiguously from the set A of all possible actions, so that the person of the physician is exchangeable in this process? Or formulated in a different fashion: Is there for the set $D = D_{t_1}, D_{t_2}, \ldots ., D_{t_n}$ of the data-sets at $t_1, \ldots ., t_n$ concerning the patient and the potential set Pot(A) of the set A of all possible actions, a portrayal $v: D \rightarrow$ Pot(A), that will render the clinical process a rule-governed behavior and unambiguously provide the physician in all possible clinical situations with an optimal guide for his decisions? ([22], p. 77)

The goal presented is one of objectivity in the sense of intersubjectivity. The peculiarities of the particular clinician should be eliminated towards the end of reproducible and reliable clinical judgment. As a consequence, the issue of giving a careful account of the enterprise of clinical judgment has been entered upon in earnest. Even though one might hold that some of the goals set by Sadegh-zadeh are excessively high, one should probably seek their partial realization. The essays in this volume explore different elements of this goal.

This volume exists in response to interests of both physicians and philosophers in a better understanding of clinical judgment. Because the issues raised by this endeavor are so wide-ranging, the contributors to the volume include not only philosophers but health care practitioners, as well as psychologists and individuals concerned with the logical analysis of clinical judgment. As a consequence, a multi-dimensional appraisal of clinical judgment is offered. The first section provides assessments of the roles of intuitions, hunches, and rules for reasoning in clinical judgment. Michael Scriven begins by addressing the proposition that clinical judgment, in the sense of intuitive judgment, is generally inferior to judgments made on the basis of a formula or recipe for reasoning. "Hence there are at the moment, to the best of my knowledge, no cases in which any *group* of experts has been able to outperform a formula, where enough study has been done to make it possible to construct a formula." ([24], p. 6). The lack of response to this fact Scriven takes to be grounds for the moral indictment of clinicians who could perform better were they to enlist psychologists in helping them learn, and helping medical students learn, explicit formulae for clinical decision-making. By this Scriven does not mean to assert that formulae can be advanced that will reproduce the performance of truly exceptional clinicians. However, formulae of high reliability should be produceable which would be useful in many circumstances. From this he derives an important recommendation for medical education. Since many clinical judgments are never clearly shown to be right or wrong by subsequent events, education could be maximized by the selection of special, pedagogically useful cases. For example, "training clinical judgment, if done seriously, would involve the selection and use of audio-visual material, with accompanying documentation and with built-in self-testing . . ." ([24], p. 10). In short, Scriven is suggesting that one could improve clinical judgment by recognizing that it is not an intuitive faculty but is rather based upon an identifiable set of perceptive and cognitive skills, which can be identified and trained by exposure to large numbers of carefully sequenced and categorized cases ([24], pp. 9–10).

In his commentary Elstein emphasizes many of the points made by Scriven. In particular, he indicates that one of the difficulties of clinical judgment is the distortion of data by the clinician. For example, clinicians tend to discount evidence that fails to confirm their favorite hypotheses. Moreover, clinicians collect redundant clues and act on information with low validity. As Elstein indicates, this is also a serious moral problem when the collection of such redundant clues requires invasive procedures or is dangerous or costly. In short, there is an issue of maximizing the efficiency of clinical judgments. In the journey from data collection through hypothesis generation, cue interpretation, and hypothesis evaluation, to clinical decision-making, there will then be many junctures at which clinical judgment can be augmented by a careful, analytic approach. For example,

> assistance to intuitive clinical inference can be provided by actuarial methods that incorporate frequency estimates or cue-disease correlations based on larger samples than would be possible for the average clinician to accumulate and without the distortions or simplifications of memory that are inevitably part of the repertory of clinical experience. ([6], p. 27)

The intuitions of the clinician can be usefully augmented.

The question is, what would such augmentation do to medicine as an art. On this point Elliott Sober approaches the characteristics of clinical judgment under the rubric of an analysis of clinical judgment as an art. An extreme (and distorting) interpretation of clinical judgment as an art would, as Sober shows, portray the clinician as achieving an intuitive insight that is inherently non-logical, which deals with unique objects (e.g., patients) in circumstances essentially marked by human emotion, and which is qualitative in its deliverances and not quantitative. Such views of clinical judgment are, as he indicates, based on false dichotomies between intuition and logic, the unique and the general, emotion and reason, the qualitative and the quantitative.

> The idea that clinical judgment involves some kind of nonlogical intuitive insight . . . has no more evidence on its side than the fact that we are now very much in the dark about how human beings solve problems. But ignorance of the logic of clinical discovery is no reason to think that there is no such logic. With respect to the other alleged dichotomies . . . the uniqueness of the individual, the role of the emotions, and the use of qualitative concepts [can be understood] in such a way that clinical judgment can be seen as a scientific undertaking. ([25], p. 37)

Once clinical judgment is no longer viewed as an unanalysable art but rather as a set of skills in problem-solving like our ordinary everyday skills in problem-solving, one can reasonably study the social context and psychological mechanisms, as well as logical presuppositions, involved in the reasoning of individual clinicians. Clinical judgment can then be analysed in the light of the psychology of problem-solving and logical models of how rational inference best proceeds. It does not, however, follow from this that the best way to learn to be a skilled clinician is merely to study some explicit and detailed specification of rules of inference. That would be as mistaken as to think that one learns to speak a language by studying its grammar alone. However, as with the acquisition of any language, it is important and helpful to know the essential characteristics of the grammar of good clinical judgments. Moreover, even in the presence of such rules, clinical judgment remains in a very important sense an art, just as being a good chess player remains an art, even though computer programming can be provided for chess. Studying rules for playing chess well can be instructive for someone wishing to be a better player. Yet much learning continues to be acquired through practice.

> In learning how to speak a language, learning how to play chess, and learning how to make clinical diagnoses, one absorbs the theory indirectly by participating in the practice on which the theory is based. It is this fact about clinical diagnosis, and not its allegedly nonlogical, nongeneral, emotional, or qualitative characteristics, that makes it an art. ([25], pp. 42–43)

Still, as Rosenhan indicates, explicit attention to the steps and elements of clinical judgment (e.g., nosological consistency, nosological adequacy, etc.) has a profound impact on the reliability, usefulness, and verifiability of diagnoses. To recur to the analogy with language, in order to speak well one must have a good command of idiom, vocabulary and grammar.

The second section of the volume is composed of a series of papers which addresses the logic of reasoning in the health care professions. As the commentaries show, the papers evoked spirited discussion and analysis of issues. In his paper, 'Classification and its Alternatives,' Edmond Murphy addresses basic issues in the characterization of medical data. How, for example, is one to sort patients into groups – as male or female, as hypertensive or non-hypertensive? As Murphy shows, attempts to classify are rarely univocal and lead to misjudgments if adhered to in a wooden fashion. For instance, no single criterion for sorting humans into males and females will satisfy all our general everyday intuitions about how females and males are to be distinguished. No single criterion (e.g., no particular diastolic and/or systolic arterial pressure levels) will be sufficient to sort all individuals into one group needing treatment for hypertension and another group not in need of therapy. Any simple approach to classification will run into difficulties because of problems in distinguishing true data (i.e., message) from erroneous but distracting data (i.e., noise), and because of the difficulties of distinguishing true from misleading correlations. To indicate some of the problems, Murphy provides a discussion of difficulties associated with regression analysis and the identification of clusters of data. In commentary Morton Beckner indicates that there is no alternative to classification. There are, rather, alternative methods of classification. In particular circumstances, certain forms of classification recommend themselves over others. As a result, one must always look at the way a particular concept functions in a classificatory scheme. Thus, even if a particular blood-pressure level were used to identify someone as hypertensive, it would be the fault of the clinician (i.e., a deterioration in the information content of the message) if the physician treating forgot *how* hypertensive the patient in fact was, and simply treated the patient as hypertensive without qualification (i.e., used the classification "hypertensive" in a simple, binary fashion – you have it or you have not).

These issues become explicit in attempts to simulate clinical judgment. In his essay addressing what John Gedye terms 'technological psychology', he engages many of the problems of classification – for example, what threshold of what measurement should be used as a criterion for safely asserting that an individual has or does not have a particular abnormality. Gedye pursues such issues by asking how one could simulate actual clinical judgment – he has in mind here simulation in the form of artificial intelligence, computers. In fact, the computer can reveal characteristics of clinical judgment and help augment the abilities of clinicians.

> The computer, by enhancing the clinician's ability to monitor his own activity in the circumstances of the high rate of turnover of clinical cases, may allow more explicit application of the judicial processes . . . , thus enabling him to commit himself more wholeheartedly to a continual activity of preparation for the next case, an ongoing rehearsal entered into in the knowledge that the curtain may at any moment rise, unannounced, on a live performance. ([8], p. 108)

As Ernan McMullin indicates in commentary, computer simulation of clinical judgment need not imply repetition of the same steps in reasoning that actual clinicians use.

> The *real* point of Gedye's proposal is not that [the computer] simulates what clinicians actually do, but rather that it suggests a program whereby clinical diagnosis could be made considerably more effective, quite independent of the question as to whether the program represents what clinicians are already doing and simply does it a little better. . . . Properly programmed, [the computer] may enable the clinician to do something he was *not* doing before, namely to make numerical estimates based on explicit and exact correlations over an extensive material. ([15], p. 122)

Even given these limitations, the computer can, as Gedye remarks in his response to McMullin, still make an important contribution by helping us simulate the clinician at his or her best.

The project of constructing programs for clinical judgment raises questions as to how one should provide reconstructions of clinical reasoning. Suppes in his contribution addresses the virtues of Bayesian versus other approaches to making medical decisions. The difficulties with the Bayesian approach are, as he notes, numerous: 1) many physicians are unwilling to use probability judgments explicitly in diagnosing the illness of patients (obviously a difficulty with any similar approach to clinical judgment); 2) the reduction of the criteria for clinical judgment to explicit rules or recognition procedures often turns out to be unworkable or extremely difficult; 3) there is no natural, non-arbitrary way to incorporate new objective evidence within a feasible application of Bayes' theory ("we especially do not have such rules when we suspect there is strong probabilistic dependence among various parts of the evidence being considered." [26], p. 148). Given the present state of the art, what seems desirable is the development of an actuarial approach to forming diagnoses and making treatment choices. Such goals should be pursued while bearing the following points in mind: 1) it will never be possible to have completely reliable conclusions about any given patient; 2) the process of clinical judgment can be studied even if it is inaccessible to experimental manipulation; 3) even crude approximations to the joint distribution of causes or symptoms through the analysis of clinical data would be a methodological step forward; 4) consideration of alternative hypotheses can make an important contribution. Such goals, however, do not commit one, as Martin Lean points out, to a Bayesian approach. The moral is a general one: "a Bayesian (or other) statistical approach could contribute greatly to accurate diagnosis, to prediction of prediction of medical results, and thus to more informed decisions as to treatment alternatives." ([13], p. 164).

The final section of major papers addresses clinical judgment from the perspective of clinicians with special attention to the influence of the clinical situation and its goals upon the decisions of health care practitioners. As Edmund Pellegrino demonstrates, clinical judgment is in the end goal-oriented. It is aimed at clinical action, at restoration, healing, and prevention, at addressing the state of affairs of a particular patient. As a result, judgments as to what is best to do turn on a nexus of value judgments, including those of the patient and the physician. Pellegrino suggests that the quandaries of the clinical context can be reduced to three basic questions. The first is "What can be wrong?" Unfortunately, clinical reality is usually such that the data available are unstandardized, poorly quantified, and often unreliable and inaccurate. In short, much of what has been said above about improving the data bases of physicians is pertinent here. The reality remains,however, one of making do with insufficient information. The same problems hamper answering the second question, "What can be done?" As Pellegrino indicates, genuinely scientific information in therapeutics is scanty. "One has only to recall the sad history of anticoagulants in myocardial infarction, the recent national dilemma about influenza vaccination, or the belated appreciation of the long-term vascular effects of oral contraceptives or antidiabetic agents." ([18], p. 178). The third question is, however, the most vexing. "What *should* be done for this patient?" An answer here can be given only on the basis of the patient's and the physician's value choices. It is here that not only ethics and value theory play a large role, but rhetoric as well – the physician and the patient attempting to persuade each other toward particular courses of action. Clinical judgment is, in short, complex. This is the case, as Pellegrino underlines, because medicine involves not only a science and an art, but the virtue of prudence. One must, with prudence, decide what one ought to do through strategies of coping with incomplete data. One must decide on the relative costs of false positive and false negative diagnoses while acting to rule out some serious conditions and venture therapeutic trials for others, because the data are never all in. And of course, one must weigh in the equation the costs in morbidity and money of diagnostic intervention and therapeutic ventures. Unfortunately, about this element of clinical prudence we know very little, and as Pellegrino indicates, we need to know more. For example, is it better or worse, and in what sense, and to what degree, to be a therapeutic enthusiast and treat certain conditions at the slightest indication for intervention? Finally, to decide what is a good intervention will turn upon ethical and other values and the negotiations of patients and physicians – a nexus which inextricably binds the practice of medicine to the humanities. This nexus is, as Marjorie Grene indicates, often to be described only through Polanyi's account of tacit knowing, of knowledge "as relying-on-clues-in-order-to-attend-to-what-they-are-clues-to." ([9], p. 197).

Such assessments of what could be loosely termed 'subjective information' receive careful address in the essay of Eric Cassell. There, among other issues, he compares the significance of "objective" versus "subjective" information about patients as such information is characterized in, for example, the problem-oriented medical record [28]. He examines the ways in which patients as psychologic and sociologic persons are viewed and treated (with attendant distortions) as objective things. In this regard, Cassell warns that the clinician must not distort the subjective through inappropriate objectifications. He or she must attend to how *particular* patients assign values, interpret causality, and understand the relationship between their current illness and past experiences. Such assessments of the patient will require viewing the patient as playing multiple simultaneous roles with him or herself. "The assigner of understanding (within the patient) takes a history from the experiencer (within the patient) in much the same manner that the physician questions the patient." ([4], p. 15). Thus the clinician must know how the patient interprets the words the clinician uses in explaining a disease to the patient. However, as Daniel Wikler points out, one will have to attend to the various senses of 'subjective' in order to come to terms with these important issues. Cassell addresses the subjective in the senses of mere beliefs and hunches, of facts about the self (i.e., private events) and of expressions of attitude, feeling, or taste. The clinician is interested in the subjective context of the patient in that the clinician is not concerned simply with whether what the patient believes is actually true, but also because that patient believes it. How a patient sees his or her illness is inextricably part of what must be taken into account in addressing that illness. These subjective issues must be approached, as Wikler shows, with as much general objective, psychological, and sociological data as possible.

The final section brings together in a series of concluding discussions a number of reflections on the nature of clinical judgment and its implications for health care. Stuart Spicker, for example, discusses how clinical judgment and choices are always made within a particular cultural milieu. In our case, it is a milieu structured by a value system which favors treatment over benign neglect. We act with reference to a clinical imperative which urges the health care practitioner to treat when in doubt. Yet, on the other hand, when decisions not to treat are embraced, they are often made without consulting the patient. The actual structure of clinical judgments is thus strongly influenced by the values and other assumptions of the cultural milieu in which they are fashioned. This point is emphasized by James Potchen et al. in their analysis of how data are used in clinical judgment-making, and how the value system of the clinician and the patient has an impact upon the discrepancies between clinical judgment and diagnostic reality. This point is also developed by Sally Gadow, who in an analysis of the implications of the theories of Foucault and Hegel discusses the role of the "whole person" of the patient in forming adequate clinical judgments. In doing so she points out the different roles played by essentialist views of disease entities, empirical medicine (i.e., patho-anatomical medicine) and probabilistic medicine in developing an adequate description of the sick individual. In stressing the person as the final synthesis of these three moments, she joins with Thomas Hill, who stresses the inextricable role of value assumptions in clinical judgment. Given this rich and complex nature of clinical judgment, it is, as Bernard Towers points out, more than can be adequately addressed through the multiple choice exams of much of modern medical education. Its complexity suggests that modern medicine is failing in the task of adequately teaching its students to be good clinicians. As the concluding remarks show, these difficulties turn on root conceptual issues that have a long history in medicine, and which this volume only begins to address.

As these essays demonstrate, an adequate assessment of the significance of clinical judgment is a complex endeavor in the epistemology of medicine. It involves, as Bieganski suggested, a fundamental critique of medical knowledge and the methods of clinical knowing. It is, moreover, much more than simply an enterprise in epistemology. Clinical judgments in their rich and full sense are freighted with values, including ethical and moral values. Evaluation and explanation properly and inextricably are bound together in medicine in general and in clinical judgment in particular – as each of the volumes of this series has shown. Ideas, concepts, and notional presuppositions and structures, including value judgments, fashion the actual practice of medicine. Medicine more than most endeavors of knowledge and technology is involved in the entire range of human values and the whole gamut of levels of reality (i.e., from subcellular processes to psychological and sociological interactions). In studying medical knowledge, in analyzing clinical judgment, one thus addresses a first instance of knowing and doing, the better comprehension of which is likely to illuminate our understanding of science and technology in general. That is, understanding medicine may shed light on areas of science and technology where the interplay of facts and values may not be as salient, or the consequences of different views of science and technology as immediately or as intimately intrusive. The analysis of clinical judgment is, as these essays have shown, far from a parochial endeavor. Such analysis turns on general issues of how complex knowledge claims are fashioned and founded, on how one classifies the elements of appearance, and on how evaluation of the objects and outcomes of knowledge claims plays a role in such knowledge claims themselves. This first volume in the series explicitly devoted to clinical judgment is offered as an initial step towards a better understanding of the multiple and intricate issues.

I wish to close by expressing gratitude to Bernard Towers for his aid in taking this step. His contributions of ideas and energies were indispensable.

April, 1978

NOTE

[1] It is not clear from the text whether the introduction to Bieganski's work ([1]) is that of the author or written by the translator, A. Fabin.

BIBLIOGRAPHY

1. Bieganski, Wladyslaw: 1894, *Logika Medycyny*, Kowalewski, Warsaw. 1909: *Medizinische Logik*, trans. by A. Fabin, C. Kabitzsch, Wurzburg.
2. Bjorn, John, and Cross, Harold: 1970, *The Problem-Oriented Private Practice of Medicine*, Modern Hospital Publications, Chicago.
3. Blane, Gilbert: 1822, *Elements of Medical Logic*, Huntington and Hopkins, Hartford.
4. Cassell, Eric: 1979, 'The Subjective in Clinical Judgment,' in this volume, pp. 199–215.
5. Cullen, William: 1769, *Synopsis Nosologiae Methodicae*, Edinburgh.
6. Elstein, Arthur: 1979, 'Human Factors in Clinical Judgment: Discussion of Scriven's "Clinical Judgment", ' in this volume, pp. 17–28.
7. Feinstein, Alvan R.: 1967, *Clinical Judgment*, Krieger, Huntington, New York.
8. Gedye, John L.: 1979, 'Simulating Clinical Judgment (1): An Essay in Technological Psychology,' in this volume, pp. 131–134.

9. Grene, Majorie: 1979, 'Comments on Pellegrino, "Anatomy of Clinical Judgment",' in this volume, pp. 195–197.
10. Gross, Rudolf: 1976, *Zur klinischen Dimension der Medizin*, ed. by Paul Lüth, Hippokrates, Stuttgart.
11. Hippocrates: 1923, *Epidemics*, trans. by W. H. S. Jones, Wm. Heinemann, London, p. 165.
12. Hippocrates: 1923, *Precepts*, trans. by W. H. S. Jones, Wm. Heinemann, London, p. 313.
13. Lean, Martin E.: 1979, 'Suppes on the Logic of Clinical Judgment,' in this volume, pp. 161–166.
14. Lusted, Lee B.: 1968, *Introduction to Medical Decision Making*, Charles C. Thomas, Springfield, Illinois.
15. McMullin, Ernan: 1979, 'A Clinician's Quest for Certainty,' in this volume, pp. 115–129.
16. Murphy, Edmond A.: 1976, *The Logic of Medicine*, Johns Hopkins, Baltimore.
17. Oesterlen, F.: 1852, *Medizinische Logik*, H. Laupp, Tübingen. 1855, *Medical Logic*, ed. and trans. by G. Whitley, C. Kabitzsch, Wurzburg.
18. Pellegrino, Edmund: 1979, 'The Anatomy of Clinical Judgments: Some Notes on Right Reason and Right Action,' in this volume, pp. 169–194.
19. Polanyi, Michael: 1967, *The Tacit Dimension*, Doubleday Anchor, Garden City, New York.
20. Proppe, A.: 1970, 'Notwendigkeit und Problematik einer Computer-Diagnostik,' in *Computer: Werkzeug der Medizin*, ed. by C. Ehlers, N. Hollberg, and A. Proppe, Berlin. My translation.
21. Rothschuh, Karl E.: 1965, *Prinzipien der Medizin*, Urban and Schwarzenberg, Munich.
22. Sadegh-zadeh, Kazem: 1977, 'Grundlagenprobleme einer Theorie der klinischen Praxis,' *Metamed* **1** (March), 76–102.
23. Sauvages, Francois Boissier de: 1768, *Methodica sistens Morborum Classes*, Fratrum de Tournes, Amsterdam.
24. Scriven, Michael: 1979, 'Clinical Judgment,' in this volume, pp. 3–16.
25. Sober, Elliott: 1979, 'The Art and Science of Clinical Judgment: An Informational Approach,' in this volume, pp. 29–44.
26. Suppes, Patrick: 1979, 'The Logic of Clinical Judgment: Bayesian and Other Approaches,' in this volume, pp. 145–159.
27. Sydenham, Thomas: 1848, *The Works of Thomas Sydenham*, 3rd ed., trans. by Greenhill, The Sydenham Society, London.
28. Weed, Lawrence: 1969, *Medical Records, Medical Education, and Patient Care*, Case Western Reserve Press, Cleveland.
29. Westmeyer, Hans: 1972, *Logik der Diagnostik – Grundlagen einer normativen Diagnostik*, Kohlhammer, Stuttgart.
30. Wieland, Wolfgang: 1975, *Diagnose: Überlegungen zur Medizintheorie*, Walter de Gruyter, Berlin.
31. Wulff, Henrik R.: 1976, *Rational Diagnosis and Treatment*, Blackwell, Oxford.

SHERMAN M. MELLINKOFF

PROLOGUE TO THE SYMPOSIUM

It is my pleasure to greet this array of scholars for the Fifth Trans-Disciplinary Symposium on Philosophy and Medicine. We rejoice in this gathering of teachers from such disparate fields as law, philosophy, history and medicine as a celebration of the boundless spirit of inquiry which is at the heart of universities. Without this spirit civilization would surely perish. We celebrate also the integral place of medical schools within universities, where they have been more or less securely and intermittently in the past, and where they seem at present happily destined to remain.

The subject of this Symposium, 'Clinical Judgment', may well have preceded history. For example, it seems reasonable to suppose that some Stone Age mother, like my own grandmother, uncannily perceived the first manifestations of acute infectious disease in the expression of a child's eyes. In the Second Book of Kings the encounter between Naaman, thought to be a leper, and the prophet, Elisha, may in fact have exemplified Elisha's clinical judgment that Naaman suffered from "desert sores", responsive to daily baths, rather than from the still dreaded Hansen's disease. [2]

Hippocrates was the first in this domain to turn the mind's eye inward, as it were, not only to recommend techniques for sharpening clinical judgment and cumulatively learning from it, but also to recognize its frailties and the predicament of the man who must rely upon it. "Life is short; the art long, opportunity fleeting; experiment treacherous, and judgment difficult." [1]

Twenty-five centuries later, holding in our hands the fruits of unnumbered gifted clinicians and scientists, we still share the dilemma as Hippocrates posed it: judgment is required before all the relevant truth can be attained.

In some respects modern clinical judgment has been rendered even more difficult, despite tremendous scientific advances, especially in the last half century. There are now sometimes judgments in the penumbra between life and death. Is a particular decision more in the light or more in the eternal darkness, and how does this decision affect what is wise and ethical for us to do?

Small wonder, then, that physicians turn to colleagues in other branches of our common progenitor, philosophy, to help us ponder questions so inseparable from human life that they have only multiplied with the acquisition of knowledge. It is thus with a renewed sense of kinship that I cordially welcome you all on behalf of the UCLA School of Medicine.

University of California at Los Angeles, School of Medicine
Los Angeles, California

BIBLIOGRAPHY

1. Hippocrates, 'Aphorisms,' I, 1 in Jones, W. H. S. (trans,), Harvard University Press, Cambridge, p. 99.
2. Kings II, 5, 1–27.

SECTION I

INTUITIONS, HUNCHES, AND RULES FOR REASONING

MICHAEL SCRIVEN

CLINICAL JUDGMENT

INTRODUCTION

The task of an opening paper in a volume is twofold; to lay out at least some of the issues and to avoid boring the reader. People cleverer than I achieve the latter at conferences by telling jokes. Here I shall endeavor to substitute outrage for amusement. Nothing keeps one awake like nursing an outrage. As to the task of surveying the issues, I shall do that relatively briefly since my fellow contributor to this section, Dr. Elstein, will be saying something about that too. He does it very well, as those who know his article in *Science* [4] will acknowledge.

THE BACKGROUND

The two most interesting facts about clinical judgment are that it does sometimes turn out to be right even when the judge cannot explicate his or her reasons for the judgment; and that, on the other hand, it is virtually always *less* reliable than a simple formula relating *some* relevant antecedent variables to the criterion variable, *when* that formula has been empirically fitted to previous data. Each of the qualifications in these assertions is important, especially in the latter. The formula is *simple* because it is linear; it could hardly be simpler. It does not pretend to include *all* the relevant variables, but only a few that are empirically discovered to contribute significantly to the reliability of the prognosis or diagnosis – usually fewer variables than the clinician believes to be relevant and for which s/he knows the values. The claim made is only relevant to tasks where we have enough background data to have fitted a linear regression model; it will not, for example, generate a conclusion with regard to the interpretation of novel dream content (see [6]), or with regard to a tropical disease where we have no track records of etiology or development or have them but have not done the work of identifying variables and fitting the coefficient. Nevertheless, the situation is enormously threatening to most clinicians for it means that they are in some sense rather easily obsoleted as engines for generating useful knowledge. Even if simple formulae will not do it, there is the lurking threat that it might be done with

H. T. Engelhardt, Jr., S. F. Spicker and B. Towers (eds.), Clinical Judgment: A Critical Appraisal, 3–16.

formulae incorporating interactions or quadratic form, without which very little of physics would be possible but with which most physical phenomena can be at least approximately analyzed.

As Thorne puts it in his *Clinical Judgment* [11], referring to the publication of the facts just described in Meehl's classic work *Clinical vs. Statistical Prediction* [6]:

> Reactions of insecurity became evident everywhere . . . There arose an immediate danger that clinical judgments would be relegated to Limbo in preference for the greater scientific respectability of objective measurements ([11], p. x).

It was not long before a barrage of rationalizations emerged, either as criticisms of Meehl's work, or as attempts to identify other areas in which clinical judgment could reasonably be expected to be superior to the clerk or the computer. The quality of these efforts has been uniformly though not exceptionlessly negligible. Meehl had not only made the strongest case against the validity of (most) clinical judgment, he had also made the strongest case *for* it that could be made in the light of the evidence.

The situation with regard to the clinician is not unlike that in certain other areas of apparent expertise. Studies of the placebo effect and of psychotherapy and of surgical procedures like pneumothorax have flattened mountains of pretentious claims for medical marvels. Research on the Hawthorne effect has done the same in many areas of educational innovation or personnel practices. Proper care in the study of moral education, or of management techniques, or teaching styles, has disclosed the virtual absence of *any* significant superiority in *any* of the highly touted and professionally endorsed practices. The greatest achievements of scientific method in the twentieth century include a large number of demonstrations that accepted scientific or expert practices are, not to put too fine a point on it, bogus.

1. FOREGROUND

It is our task in this session and at this conference to see exactly what sound structures can be or have been built from these ruins. I shall begin by reviewing in a sentence or two one of the most brilliant series of studies that has occurred since the publication of Meehl's book in 1954, though it goes back to a lone piece of masterly research much earlier, the study of "what goes on in the corn judge's head". In this series of studies several attempts were made to model and thence to improve on the information-processing function operating in the clinician's head. Naturally, the first source of suggestions was the clinician, and what Meehl is telling us is that the model the clinician *actually* uses – whatever that is – is less valid than a linear regression equation fitted to the same data. What has turned up since is that increasingly cynical views of what it takes to be a winning formula are increasingly more accurate. In particular, it turned out that using linear equations with *randomly* assigned co-efficients will, on the average, do better than the clinician. And it then turned out that the chore of generating random numbers for the coefficients and averaging the results could also be dispensed with because making the coefficients all unity produced a formula which beat both the clinician and the random formula.[1] The ultimate ignominy is surely to discover that the vast experience and formal training of the clinician results in judgments no better than the simplest possible formula with the simplest possible coefficients.

Of course, there are, as before, some conditions on this result that somewhat limit its significance. But notice first that the result is in some respects more general than the earlier one. In particular, we do not require the detailed data for a "fitting" operation that was previously required; however, we do need to know that the "predictor" variables we are using *are* positively or negatively correlated with the dependent variable over the relevant range in order to give them a +1 or –1 coefficient. Not very much to ask, to be sure, but it is the key to the whole issue in a way because if one asks the crucial question *why* the clinician does so badly, it is partly and perhaps chiefly because s/he gives weight to variables that do *not* have this simple property. It is *also* because his or her head uses a formula which attributes interactions where they do not exist, and relative and variable weightings that are not justified. No one, I believe, has yet identified the exact proportion of the error that is due to each of these causes, but that each contributes substantially I think there can be little doubt.

Four footnotes should be added to the above picture. First, a case recently emerged in which it appeared that someone had uncovered an exception to the general inferiority of "clinical" i.e., intuitive judgment. In a study whose relevance to the medical-psychological scene will not be doubted by those familiar with the extreme constancy of the phenomenon across all fields of expert judgment, Libby discovered that bank officers predicting bankruptcy within three years were indeed better than the formula – but only by .9% at the 44% level. Moreover, as we suggested earlier when it turned out that they were using financial ratios, then by using a *single* ratio (assets/liabilities), Goldberg [5] showed on reanalysis that one could predict 80% correctly, 36% better than the experts. That is, a step up from the restriction to the linear formula immediately reversed the only exceptional case – and that one trivial if one even breathes the word cost-effectiveness into the atmosphere of the comparison.[2]

Second footnote: the one or two apparent exceptions to the general picture that were reported in *Clinical vs. Statistical Pronunciation* did not survive replication or reanalysis. Hence there are at the moment, to the best of my knowledge, no cases in which any *group* of experts has been able to outperform a formula, where enough study has been done to make it possible to construct a formula.

Third footnote: the discovery that level weighting of the variables is the best strategy when one lacks the data for exact fitting, and is significantly better than expert judges, will remind many of the analogous and in fact supportive result in the theory of tests and measurement concerning the differential weighting of sub-tests when calculating an optimal score from which to predict a criterion variable. We find there that variations from level weighting, which seem so obviously important on a priori grounds, in fact make very little difference in practice.[3] An obvious partial reason is that high correlations between the scores of individuals on the sub-tests, a quite common situation, will mean that changes in the weights will not affect the rankings and hence not affect the positions. But the result applies rather well even when rather low correlations exist on the sub-scores. For if the condition is met that was stated earlier with respect to the claim that level weighting is a giant-killer, namely that the variables used are significantly correlated with the criterion variable, then they will be mildly correlated with each other and the inevitable consequence from the sub-test studies applies.

Fourth footnote: there is one application of this kind of result that one should hesitate to make, though not for scientific reasons in the usual sense. The result should not be applied cavalierly to *normative* or *evaluative* formulae, especially if this application is known in advance. Suppose you rate faculty on their teaching, research, campus service and public service – four variables. How should you weight them? If "the formula result" is applied retrospectively, you might as well – as validly – rate them equally as get into arguments about how to justify differential weighting. But if this formula is to be announced in advance of its applications, as justice typically requires, then it will *not* do to level weight the factors since the formula will then be open to exploitative or manipulative behavior. The validity of level weighting is not a mathematical but an empirical result and amongst empirical results it is not a law of nature but a *typical pattern* of behavior like the tendency of the very young to trust grown-ups, and is hence open to exploitation. This point is not so distant from our subject today if you reflect upon the power of *prescriptions* (i.e., normative recommendations) in psychotherapy.

2. THE FUTURE

In this oppressive atmosphere, as it has seemed to many clinicians, there are of course some rays of sunshine. The most obvious one, already mentioned, is the fact that there *is* no formula in most cases. However, that is principally because of lack of diligency by medical clinicians in getting psychologists – who have the necessary skills but lack the status – to develop them. It is thus of some moral note, but of little scientific interest, since it represents nothing that by its nature must, nor on moral grounds should, persist.

The most interesting special case is perhaps the clinician's role in generating particular hypotheses. Meehl's example of a particular dream-interpretation is impressive. It would be still more impressive if the frequency, validity and therapeutic relevance of such examples were higher. Nevertheless, this kind of example clearly by its nature transcends any formula. Thorne's own confidence about the future is inspiring. After warning us of the need for a very critical approach to the methods we do use, he says:

> However, when the issues have been completely objectified and studied, it may be assumed that completely trustworthy validated methods will eventuate. ([11], p. 4).

and again:

> The position of this book is that clinical judgment potentially is the most valuable tool available to the clinician, particularly in situations where the human element is essential. ([11], p. 4).

What I like about these quotations is that they are readily and plausibly defensible simply by construing them as tautologies. In the first quote, if "completely objectified and studied" is taken rather literally, then it *definitionally* includes the discovery of the underlying causal relationships which of course would provide us, in principle at least, with "completely trustworthy validated methods". And in the second quotation, it is clear that in "situations where the human element is essential" the "most valuable tool available" will indeed be clinical judgment since it will be the *only* tool available. If it were not the only tool, then the human element would not be essential.

I find little solace in tautologies. To get a little tougher about the situation, consider the analogy to clinical judgment in the experienced blackjack player. Such players have a good feel for the game and certainly lose less money than beginners. Now blackjack happens to be a genre where we *have* "completely objectified and studied" the issues, and we have discovered "completely trustworthy validated methods", in Thorne's words. But there is not a player in a million who uses those methods, despite the fact that they and – as far as we know – they alone actually win against the house over long runs. It is a touching supposition that clinicians in the health fields will somehow take advantage of validated measures whereas equally intelligent gamblers with at least as many hours of experience, and *with much stronger reasons* for using the methods, will not.

On this point about motivation, it is essential to see that the cultural environment and the social controls on the clinician are simply not putting serious pressure on her or him to be right. The occasional professional with a real drive to cut through the verbiage and the status games of case conferences or convention symposia emerges – indeed Thorne and Meehl have both served in that role – but the peer pressures and general methodological incompetence of most practitioners soon turns most others into herdmembers and the renegades into lone crusaders or other fields. I do not suggest for a moment that philosophers, or this philosopher, are in any way superior. I do say that the situation in medical research, most notably in the psychiatric and psychological area where the arguments about clinical judgment are most often to be heard, is not one to provide a shred of support for the views that we will either move quickly towards discovering improved methods of clinical judgment, or use the improvements when we get them. There could not be any rewards for being right because no one keeps score. We all have anecdotes about brilliant examples of clinical insight, but which of us has a logbook with validated counts on our own or anyone else's long-term track record? There are fifty of those log-books to be found on the floor of the big clubs in Vegas any night of the week.

Their absence from the clinical field is the give-away indicator of the value-system there. To put it another way, it is about a quarter-century since we discovered that a formula can trounce a physician if you're prepared to do a little work identifying relevant variables. How many such formulae have been produced by the prodigal colossus of medical research since then? How many systematic *efforts* have been made to find them? Silence speaks louder than words.

The lack of motivation and the lack of the necessary data-systems and feedback loops are not the only troubles. When one discovers that the way to win at blackjack requires keeping a mental count of all the cards that have been dealt, in an unobtrusive way, we rapidly find that a large slice of our "experts" are up against the IQ barrier. This is not a skill they can manage. There is not the least reason to suppose that the hypothetical techniques of clinical judgment that might turn the tables on the statistician will be techniques that most clinicians can *manage*, even if they were strongly motivated to do so.

Here it is instructive to reflect on an area of expertise that involves truly clinical judgment in order to succeed, where it *does* succeed, and where *no* mechanical solution will do the trick, for that is the situation clinicians suppose to hold in the medical field. Poker would be my choice, though chess and bridge are also worth discussing. These games are crucial paradigms for us, whereas the other sciences are not, because there is an objective measure of success present that not even Kuhn's scepticism can contaminate. The best poker players can skin others with equal intelligence, equal training and longer hours at the tables. They have grasped skills of perception, reaction control, synthesis, fatigue correction, rapid calculation, and speed in reading the new player that they can only partly convey in writing and talking. But they can convey enough in that way to establish (along with their performance record) the immense difficulty of achieving the skills involved in making the clinical judgments as to when a player is bluffing, lying, anxious, elated, even though trying to conceal it – truly excellent examples of clinical judgments of relevance to the professional psychologist, in industry, school, or clinic. No, I cannot share Thorne's confidence that the better methods – even if they could be found – will be useable by any significant number of the profession.

Will they or the techniques for finding them, even be part of training of clinicians? There are few signs that this is being done in a way that would maximize skill development. There is still far too much dependence on teaching theory or basic science in the medical or psychological curriculum, far too little recognition of the fact that the perceptual and cognitive skills required for clinical judgments are likely to best be identified or trained by exposure to very large numbers of carefully sequenced and categorized cases. Moving away from the psychological field for a moment, take the kind of clinical judgment a general practitioner is called on to make a dozen times a day as to the need for an X-ray of a painful hand, as to the need to immobilize a sprained ankle, as to the need for penicillin to treat a throat infection. *Most* of these judgments are never shown to be right or wrong by subsequent events: like astrological advice, they are perfectly reconcilable with any outcome. This simple fact explains the vast power of fashion in even the field of the family physician. During most of her or his training, as far as clinical experience was involved, suffered from the same handicap for the same reason – nobody runs a control group on routine hospital procedures. In addition, the *rate* of exposure to such cases on rounds for an intern is terribly low. Remember that one is trying to develop and make instantly available a truly vast repertoire of possible responses in the clinician; there are not just the hundreds of sets of diagnostic possibilities, but hundreds of sets of relevant tests, of prognoses, of treatments, of relative risks, of contingency controls, of time constraints tied to alternative developmental diagnostic and emergent crisis patterns, and so on. The learning repertoire is vast in itself; but acquiring *immediate control* over it, getting it *on-line* is an even vaster task. Training clinical judgment, if done seriously, would involve the selection and use of audio-visual material, with accompanying documentation and with built-in self-testing, that would increase the frequency of exposure by *tenfold*, and – because of the selection of cases to avoid repetition and to increase discrimination – *should* increase learning by ten-fold. But one could settle for three-fold.

Finally, in taking this hard look at the chances of improvement in the future, we have to ask ourselves about the researchers. Are *they* operating in such a way as to maximize the possible yield of improved insight into clinical judgment? The research that we have been quoting is of immense importance, but it only breaks the ground. To get new growth we need something quite different, we need to first search for and then analyze the *individual* clinicians who *are* able to outperform the formulae. In the clinical vs. statistical studies reported in Meehl [6], there was at least one such gifted individual who consistently and significantly beat not only his colleagues but the clerk with the calculator. On anecdotal grounds I am strongly inclined to think that the world of physical medicine contains others, were we to search with real diligence and due modesty instead of acting as though most clinical professors had the gift. No doubt most of them can diagnose with better than base-rate accuracy; but then most waterdiviners pick where to drill with better than chance accuracy. That doesn't show they have any gift, only that they use some cues with some validity; it turns out that the intelligent hydrologist does at least as well, and can also identify the cues so that others can learn quite quickly. The only interesting question is whether anyone is doing better than relatively simple rules can account for; we expect nothing supernatural from our clinicians, but we hope for something superstatistical whose mystery we can then *try* to unlock. (The work on the bankruptcy predictions represents the right approach in this respect.) Incidentally, the situation is just the same in the research on administrative styles: there appear to be super-administrators but no one will do work on them as individuals; it is not rewarded, it is not theoretically founded, it *looks* too limited in scope. On the contrary, it is the best bet for finding *general* results.

I am perfectly open to the possibility that we can find superclinicians whose skills we *cannot* reduce to even a *complex* formula. In the first place, I think we have hints of the existence of such geniuses in the work on the MMPI that suggested its users could perform better diagnoses from looking at the results in profile form than by looking at a list of the scores in standard order. That is, the brain's great talent lies in its skill at the pattern-recognition task, where it can discriminate the very complex interactions between visual cues, provided we make the effort to present data in visually accessible ways.

The second possibility is of course the one Meehl stresses, the ingenuity and creativity function, where a novel hypothesis is produced to explain novel configurations of data. Our capacity to generate original but appropriate and comprehensible sentences is the standing proof of the limitations of a statistical basis for modelling all our talents. There are indeed special clinicial tasks – I am thinking now of the epidemiologist acting as detective – where this skill is virtually all that counts. Who has it? Who does this better than anyone else? It would be easy to find out by preparing case-study dossiers containing the unfolding web of evidence that confronts the investigators from the National Disease Control Center, for example. This is simply a standard procedure for using criterion-referenced tests, and yet where are to be found the hundred or five hundred such dossiers in which the right answer is known to history but not learnable in advance by the candidate, being used to train new public health doctors and – for our purposes, the key use – identify the superstars?

It is more than a little disturbing that we fail to take seriously (in the sense of doing as well as we can) the training of people on whom the lives of hundreds or hundreds of thousands depend. The farce of the Legionnaires' disease investigation and its now-uncovered predecessors is probably being buried under rationalizations instead of studied as a basis for improving procedures: no doubt some little report will be done, but will every medical school in the country have incorporated the lessons to be learnt from it by next fall – will every relevant CME course have it in the pool of items for the future? What about the non-sensical reactions of the official clinicians to the discovery of paralysis in elderly people receiving the Gerald Ford Memorial flu shots? With one grand stroke they demonstrated that their capacity for diagnosis was worse than that of most intelligent laypeople in the country – or perhaps that they had sold out to political considerations. On the base-rate data available to them, which we have since seen, the null hypothesis was nailed on the first day. Yet they continued to think that "saving the program", or saving the drug companies from damage suits, was more important than saving people who were clearly endangered, whereas the threat from the flu was not clearly established.

To sum up the situation then, we are treating the possibility of gold in the brains of clinicians as if it is not worth mining, or so obviously present that it is not worth investigation, and neither situation has any support. This attitude is affecting our thinking, our training, our research and our practice and it's killing people as well as under- or over-rating clinicians.

At this point in writing the paper I turned to Dr. Elstein's excellent review and suggestions in his recent article in *Science* [4], to which I have since added a couple of implicit references earlier in this paper. I was delighted to find so much agreement, and I will just provide one or two thoughts as to areas that might bear further discussion. As a professional evaluator, I was especially pleased to see him bringing in the difference between "mere" accuracy in diagnosis etc. and accuracy weighted by consideration of the different risks involved in the case. One might add to this the importance of distinguishing the differential costs of Type 1 and Type 2 errors. I was also pleased at his stress on the psychologically different effects of sequential and simultaneous data-presentation, and on the need for much more serious attention to the determination of the actual figures (or *good* estimates) for risks and reliabilities and costs. I think he is correct in noting the gap between medical and psychological *conceptions* of clinical judgment and explaining part of it in terms of the preceding refinements and learnings that are needed on both sides. But it will be clear from my earlier remarks that I am inclined to ascribe much of the lack of communication to a lack of a suitable work environment for the working clinician, one that would lead to a serious effort to improve not just accuracy but optimality of judgment. It is only in terms of such a fundamental lack of communication that one can make sense out of the persistence of the banal and ignorant reactions that many clinicians still produce to the "threat of the formula," reactions such as that every individual is unique and hence no formulas will work, or that individual values are involved in clinical judgment and hence no general approach can be valid. These remarks are about on a par with "every athlete is unique and hence their performances cannot be compared." And there are, as Dr. Elstein rightly stresses, errors on the other side as well. As a philosopher of science I have been amazed by the cavalier way in which neo-positivist philosophers of science have dismissed the actual use by scientists of terms like explanation, predication, possibility, implication, in favor of some grotesque oversimplification that caught their fancy because of its neat mathematical properties. Mathematical models of clinical judgment have similarly oversimplified it. If I have sounded particularly critical of the clinical rather than the statistical side in the preceding, it is only because I judge that *in this case*, the clinicians have been – as a body – much more culpable than the statisticians, especially since the matter at stake affects the welfare of their patients, something less immediately apparent in disputes in the philosophy of science.

It is true that the clinicians usually have enough commitment to the welfare of patients and the search for truth, or at least to the reduction of insurance costs, to order sufficient tests and to improve radiography etc. This is working within the paradigm. But it is more than most can manage to achieve the breakthrough of reconsidering the basic methodology within which this data is processed. Elstein is excellent on the way in which threat impedes this step; again I would stress the absence of positive reinforcement for achieving it, something which the political-medical control of NIH does not facilitate. If academicians are not strongly reinforced for good teaching, they do not do it; and if clinicians are not strongly reinforced for optimal performance, *they* will not do it.

4. MAPS

To conclude, let me present two small attempted clarifications, or at least hints about the clarification, of particular aspects of the clinical judgment situation – in the spirit of what Elstein calls methodological algorithms. Perhaps in terms of the geographical analogy of my section headings, I might call them maps.

The first concerns the basic logical model of reasoning in the clinicial, jurisprudential and moral domains, domains which are not only closely analogous in this respect but which each turn up in their own right in the context of patient care procedures and facilities. Here I want to suggest two apparently contradictory views. First, in these value-laden domains, the ultimate objectivity of the value-judgments involved is in fact extremely high, as we discover if we act as though this might be so. It is a self-fulfilling claim to assert that value judgments are essentially subjective and undecidable; but to deny it even in these days of the decline of positivism requires considerable intestinal fortitude and intellectual competence. I have discussed this elsewhere in one of the Hastings Center publications on the philosophy of medicine, but stress it here because it is crucial to any understanding of clinical judgment to see it as involving an ethical dimension and not just practical reasoning, although even the latter typically involves value judgments. And since values *are* involved the degree of subjectivity in them had better not be enough to entail radical relativism about the clinical judgment itself or we are back to witch-doctoring. It is depressing to hear professional clinicians (or psychologists) acclaim the value-components' crucial importance *and* its essential subjectivity; the name for this combination, to the extent it goes beyond the doctrine of informed consent, is quackery. But conventional wisdom has elevated it to the status of unthinking acceptance and, worse, a handy defense against critical evaluation of professional work.

The other half of this point about the basic logic of clinical inference will appear to point in the other direction. For it is the assertion that the logic of clinical inference is as much weaker than statistical inference as that is weaker than the deductive inference of classical mathematics. So I am arguing both for greater strength in the foundations of clinical inference than is conventionally accepted, and greater weakness in the scaffolding. But, just as one might say paradoxically that the 'weakness' of statistical inference is in fact the source of its strength – that is, its greater utility – so the same can be said in this case. Similarly, the "weakness" of wood-frame construction in earthquake zones is *its* strength. Some of you may know something about a movement in the philosophy of law called The New Rhetoric [8]. Others will know of the work on modal and deontic logic and fuzzy set theory: or of Polanyi's work on implicit knowledge [9], or perhaps of my work on normic or normative inference [10]. Each indicates a straining at the bonds of conventional inference patterns in those areas, just as the move to Bayesian or non-parametric statistics illustrates the same within statistics itself.

Although The New Rhetoric, like Polanyi's new epistemology, doesn't work out in detail – in my opinion – it does start the process of exploring alternatives to the standard conception of scientific information. Let me call the model I am advocating the "logic of considerations (or normative logic or the logic of prima facie inference); the associated epistemology I call the theory of weak knowledge (or the epistemology of possibilities and approximations). It's a long story to tell it all, but let me just remind you of some of the "facts" of clinical or legal reasoning which such models must fit – facts in the same sense that the facts of the native speakers' intuitions are taken as facts by Chomsky [1] in developing structural linguistics. First there is the problem of the interpretation of generalizations such as

"Thou shalt not kill"
or "I before e except after c"
or "Don't castle early in the game",

or the analogous medical, legal and engineering aphorisms. It is absurd to treat these as claimed exceptionless generalizations; they are norms or prima facie truths, known to admit of many exceptions, but providing us with a simple norm to refer to, which applies *more or less* in *most* cases. It's not a statistical norm, both because it's non-quantitative and because exceptions often outnumber the confirming instances, strictly speaking. *How* can a norm be useful when *that* is the case? By being the *best simple approximation*; and this is not a refuge of the sloppy subjects, it is the way almost all "laws" in the physical sciences operate. Laws of elasticity, conductivity, gases . . . all are only approximations, not very often true at all, but very useful.

So, in clinical inference leading to clinical judgments, we operate from such rough guidelines *and these cannot be adequately formalized either as statistical or as exact generalizations*. They have a different logic, and even fuzzy set theory does not encapsulate it. The process of developing models of clinical judgment must be based on a more sophisticated set of models than most contemporary mathematicians and logicians understand.

Second, there is the problem of synthesizing relevant data in a particular case, as opposed to generalizing about a family of cases. Here is where the term "the logic of considerations" applies most clearly. What we have to synthesize are the (usually several) relevant generalizations, and the usually many relevant *estimated* values of variables, and whatever overarching theories apply, and the weightings of the several values that are involved. The optimal way to combine such diverse elements is *not* a weighted linear formula *nor* a higher order polynomial with interacting terms. These approaches are both too quantitative and too rigid to handle the way in which *primary considerations* are evaluated in the light of *overriding considerations*: That takes another kind of logic. Which is not to deny that the polynomial can't do the best job *over a certain range*; its problem is that it doesn't know when to stop.[4]

My concluding remark is just this; to have discovered that an extremely simple model of this synthesis will outperform the clinician does not show that a more sophisticated model will not outperform the linear one (cf. the bankruptcies example), *nor* does it show that clinicians cannot outperform any model. But they will never do it consistently except by climbing the ladder of the mechanical approaches and (perhaps) – *then* throwing it away.

Clinical judgment does not always involve clinical instruction. It is often purely perceptual or mainly psychotherapeutic rather than ratiocinative. But clinical performance in the medical/psychological field either inferential or perceptual, must currently be evaluated as so far inferior to that of a half-expert cardsharp, and so far from being as good as present methodology and pedagogy makes possible, as to be the main scandal of the professions in the second half of the 20th century.

University of California at Berkeley
Berkeley, California

NOTES

[1] It is of course the work of Robyn Dawes [2] and others interacting with him to which I refer.

[2] My thanks to Robyn Dawes for this update in a personal conversation.

[3] Thanks to Bob Wilson for suggesting the relationship.

[4] This line of argument is the basis of the most ingenious attack on computer simulation extant – Bert Dreyfus in *What Computers Can't Do* [3].

BIBLIOGRAPHY

1. Chomsky, N.: 1964, *Current Issues in Linguistic Theory*, Humanities Press, New Jersey.
2. Dawes, R. M.: 1967, 'How Clinical Probability Judgments May be Used to Validate Diagnostic Signs', *Journal of Clinical Psychology* **23**, 403–410.
3. Dreyfus, H. K.: 1972, *What Computers Can't Do: A Critique of Artificial Reason*, Harper & Row, New York.
4. Elstein, A. S.: 1976, 'Clinical Judgment, Psychological Research and Medical Practice', *Science* **194**, 696–700.
5. Goldberg, B. B.: 1973, *Diagnostic Ultrasound in Clinical Medicine*, Williams & Wilkins, Baltimore, Md.
6. Meehl, P. E.: 1954, *Clinical Versus Statistical Prediction: A Theoretical Analysis and a Review of the Evidence*, University of Minnesota Press, Minneapolis, Minn.
7. Meehl, P. E.: 1973, *Psychodiagnosis: Selected Papers*, University of Minnesota Press, Minneapolis, Minn.
8. Perelman, C.: 1971, *The New Rhetoric*, L. Olbrechts-Tyteca, J. Wilkinson and P. P. Weaver (trans.), University of Notre Dame Press, Notre Dame, Indiana. *See also* C. Perelman: (forthcoming), *The New Rhetoric and the Humanities*, Reidel, Dordrecht.
9. Polanyi, M.: 1966, *The Tacit Dimension*, Doubleday, New York.
10. Scriven, M.: 1976, 'The Science of Ethics', in H. T. Engelhardt, Jr. and D. Callahan (eds.), *Science Ethics and Medicine*, The Hastings Center, Hastings-on-Hudson, New York, pp. 15–43.
11. Thorne, F. C.: 1961, *Clinical Judgment, a Study of Clinical Errors*, Journal of Clinical Psychology, Brandon, Vermont.

ARTHUR S. ELSTEIN

HUMAN FACTORS IN CLINICAL JUDGMENT: DISCUSSION OF SCRIVEN'S 'CLINICAL JUDGMENT'

It would be quite impossible to comment in detail on each of the main points in Scriven's wide-ranging paper. Much of it is concerned with the issue of clinical versus statistical prediction and with the implications of the well-replicated research finding that clinical judgments can be reproduced or even improved upon by simple statistical formulas derived from a representative sample of prior judgments [11]. This appears to be a puzzling, contra-intuitive, controversial research finding. For surely its implications have been only slowly incorporated in clinical practice, and have perhaps been more resisted and avoided than attended to. This discussion, therefore, will focus on one major question: Why do simple formulas for judgment so consistently equal or exceed the accuracy of human inference? I shall answer this question in two ways: first, by offering a critique of human judgment from an information-processing perspective, and second, by critically examining the research model that produced this finding. My concern will be, then, with the psychological processes of clinical judgment and with ascertaining what research on clinical judgment has actually studied. The overall thrust of the discussion will be to agree with the main line of Scriven's argument, while pointing out that there is still an "on the other hand".

THE STUDY OF CLINICAL INFERENCE

Research on the psychology of clinical inference has been mainly carried on within one of three research paradigms: the information-processing or process-tracing approach associated with the work of DeGroot [2], Kleinmuntz [15], Newell and Simon [19], and more recently with the Michigan State group [6]; the probabilistic or Bayesian approach that views clinical inference as yet one more form of decision making under uncertainty [7, 23, 26]; and the regression or lens model approach [11, 12] that investigates capturing judgmental policy and the modeling of judgments by means of regression equations.

The information-processing view of cognition aims to characterize the underlying thought processes by recording and analyzing the steps and thoughts of clinicians as they attempt to solve clinical problems. The goal is

H. T. Engelhardt, Jr., S. F. Spicker and B. Towers (eds.), Clinical Judgment: A Critical Appraisal, 17–28.

to describe the thought process and to explain it in terms of basic psychological elements and principles. The other two approaches are concerned with how imperfect information ought to be optimally combined.

I shall draw on findings of all three research traditions to describe and explain both some strengths and weaknesses in clinical inference. I shall be concerned with showing how intuitive clinical reasoning is generally adapted to the task that must be performed and at the same time manifests some basic characteristics of the human information-processing system. Yet clinical inferences are not always correct. Mistakes can be understood within the three research paradigms mentioned. The discussion at that point will concentrate upon some pitfalls, not because they dominate clinical inference, although the regression paradigm may make it seem so, but because they could so often be avoided if formal inference systems were employed in place of intuitive clinical judgment.

Bounded Rationality

The psychological principle basic to the understanding of clinical reasoning is the concept of *bounded rationality*. This principle emphasizes that limits exist to the human capacity for rational thought which are *not* results of unconscious motives or psychodynamic conflicts. The human mind can be conceived of as a computer with certain information-processing capabilities [21, 19]. As such, it can do some things better than others and resorts to certain strategies to help overcome the limitations inherent in the system. In considering clinical reasoning, the most relevant limit is the relatively small capacity of working memory, compared to the essentially infinite size of long-term memory. This means that, in a brief time, we cannot work efficiently with all we know about a problem or all the data that could be collected. Some common features of good and poor clinical reasoning are consequences of efforts to cope with this limitation. Given the limited size of working memory, one is literally required to process data serially, to select data carefully, to represent a clinical problem in simplified ways, and to work as rationally as possible within a simplified representation. Such a representation cannot enumerate all the possibilities, but it gives the problem solver a starting place. Without it, it would be very difficult to make progress on solving a clinical problem of any significant magnitude. Nevertheless, while the principles used to simplify problems are often useful, they can lead to certain errors.

Characteristics of Clinical Reasoning

Clinical inference begins with a problem, a situation that is somehow indeterminate. The question may be, "what is wrong with this patient?" or "what are the causes of these symptoms?" or "what should we do about this situation?". These problems are resolved by a process of selectively collecting information and combining it into a judgment or decision. Unlike many situations used in experimental investigations of the psychology of thinking, all of the elements needed for a solution are not presented to the problem solver at the start of the problem, awaiting only reorganization. Some collection of data is nearly always needed. A great deal of flexibility is possible here. The clinician may choose to employ a standardized set of methods to assess all patients (the routine workup) or may adjust data collection procedures so as to illuminate the particular problems that are either presented at the start of the encounter or are discovered as work proceeds.

Another feature of clinical inference in its natural state is the flexibility of solutions available to the clinician. Diagnostic problems are solved by the hypothetico-deductive method. A limited number of hypotheses are generated very early in the clinical workup, using a very limited amount of data compared to what will eventually be collected. The clinician asks then "what findings would be observed if a particular hypothesis were true?" and proceeds to ask the questions or perform the maneuvers of the physical exam or order the laboratory tests needed to answer this question. As the workup proceeds, some early hypotheses may be dropped and new ones formulated, but experienced clinicians are often able to formulate the probable solution to a diagnostic problem as one of the early hypotheses with a surprising degree of accuracy.

This method provides a great deal of flexibility in the problem space constructed by the clinician to search for a solution and thus in the data to be collected. The set of solutions entertained by the clinician is limited by the contents of long-term memory, for clearly one cannot consider, even hypothetically, a condition one has never heard of or knows nothing about. The capacity of working memory is also a limiting factor. Empirical studies of medical reasoning [6] have shown that it is difficult to consider more than four or five hypotheses simultaneously. When that limit is reached, new candidates can be added only when old ones are dropped or (what amounts to the same thing) when an old hypothesis is reformulated into a new one. Subject only to these constraints, the clinician has great flexibility in selecting a set of four or five hypotheses out of the considerably larger number in long-term memory for evaluation in a particular problem.

One other feature is central: The method for combining information into a conclusion preferred by most clinicians is informal, intuitive or qualitative. It is not formally stated or quantitative. Clinical practitioners describe clinical judgment as an artistic, intuitive, qualitative process that is neither easily nor appropriately quantified.

These features are conveniently summarized in a four-stage model of medical inquiry. It points out that diagnostic problems are solved by an iterative process of:

(a) Data collection – the process of gathering and collecting data.

(b) Hypothesis generation – the process of generating alternative formulations of the problem.

(c) Cue interpretation – the process of interpreting the evidence collected in the light of these hypotheses.

(d) Hypothesis evaluation or judgment – the process of combining information to reach a diagnostic decision.

Therapeutic decisions flow naturally from diagnostic judgments in this model.

Clinical Inference and the Structure of Thinking

Thus far I have described a system for obtaining and processing clinical information that is reasonably effective for doing clinical work and *also* fits the specifications of an information-processing system characterized by bounded rationality. Let us see how these characteristics of clinical reasoning reflect the operations of that principle.

1. The process of gathering data is sequential, because this is the way both clinicians and patients must work. The whole story simply cannot be told at once, nor could clinicians work with all the data if it could be. Time is required to transfer portions of the data into working memory and to retrieve needed rubrics and routines from long-term memory.
2. Since time is limited and processing speed is finite, data collection is selective. The number of hypotheses considered at any moment is usually on the order of four or five and appears to have an upper bound of around six or seven. This number is clearly well below the number of hypotheses in the long-term memory store of any reasonably experienced physician.

3. The rules used for combining data are also simplified. In our studies, we found that clinicians typically interpret data on a three-point scale: as tending to confirm or disconfirm a hypothesis or as non-contributory. This weighting scheme reduces correlations, probabilities, or likelihood ratios to a much rougher, simpler breakdown than would be necessary if one were to use Bayes' theorem or regression equations to reach judgments. Furthermore, even this three-point scale was obtained under quasi-experimental conditions. Most clinicians, as I have indicated, assert that the combination rule cannot be formally stated or quantified. It is, of course, on this latter score that research on clinical judgment has most vigorously disputed the received clinical wisdom [10, 23, 5].

PROBLEMS OF CLINICAL METHOD

I want to turn now to some of the pitfalls or difficulties that such a system may encounter. Again, I stress that I have no data about how frequent these errors are, nor do I claim that clinical information processing is substantially without merit. Yet when students learn the process of clinical inference, these are some of the pitfalls most frequently encountered.

Bias Caused by Hypotheses

As we have seen, clinical diagnostic problems are characteristically solved by formulating a small number of alternative solutions and testing them. While it is possible to reformulate the problem as one moves along and to collect some data routinely, it is nonetheless the case that these preliminary formulations help to define an area within which a solution is sought and do partially guide subsequent data collection. These early formulations can be misleading. They may direct attention to an irrelevant feature of the problem, cause the clinician to engage in a search for inconsequential cues that would otherwise be ignored, or lead the decision maker to refrain from a useful search for cues that would otherwise be collected.

Moreover, the data in a clinical problem do not simply speak for themselves. Like all facts, they are filtered and interpreted through our expectations. The data best remembered tend to be those that fit the hypotheses generated. Where findings are distorted in recall or otherwise discounted, it is generally in the direction of making the facts more consistent with particular diagnostic pictures. Thus, clinical memory displays a phenomenon described long ago by Bartlett [1].

Another illustration of the bias in clinical judgment that can be produced by expectation comes from a clinical problem solving exercise in which the opening statement of the problem suggested that the patient had a herniated lumbar disc [6]. Three of the four physicians who did not reach the correct diagnostic conclusion (out of 15) made a diagnosis of herniated lumbar disc and restricted their data collection so that evidence supporting other alternatives was neglected, a very clear example of the effect of suggestion on thought processes.

It might be simpler if clinicians were to steadfastly discipline themselves not to generate early hypotheses at all and thus avoid the biases they create. For better or worse, however, it seems practically impossible to reason without hypotheses. People are invariably trying to make sense out of their experience as it unfolds and are always generating hypotheses to explain their observations. With experienced clinicians, these early expectations probably are more often helpful than deceptive. It is best to consider several hypotheses and to construct these deliberately as alternatives to the idea or ideas immediately suggested by the problem. In that way, a premature jump to an erroneous conclusion could be avoided.

Because of the limited size of working memory, it is difficult for even experienced physicians to work with more than four or five diagnostic or therapeutic alternatives at one time. One strategy for increasing the capacity of working memory is to process information sequentially by "nesting" hypotheses. Four or five possibilities of infectious disease may be considered under the general category "infection" so that at a higher level of generality more working space is temporarily available. At some point in the workup, that branch can either be discarded or more exhaustively examined.

Overemphasizing Positive Findings

Clinical findings usually bear a probabilistic relationship to the underlying causes that produce them. As clinicians know only too well, very few signs or symptoms are pathognomonic of a particular disease. Many more are associated with several diseases and diagnostic decisions are reached by somehow weighing and combining evidence. One pitfall in clinical reasoning is to discount evidence that fails to confirm one's favored hypothesis on the grounds that there is only a probabilistic relationship between evidence and hypotheses anyway and a perfect match was not to be expected. This error is equivalent to overemphasizing data affirming a hypothesis and slighting data that tend to disconfirm it.

A classic illustration of this pitfall is found in a study of clinical inference [25]. A group of nurses was presented with a series of cases in which the presence or absence of a particular symptom was associated equally often with the presence or absence of a particular diagnosis. Each of the four possible combinations occurred 25% of the time in a series of brief case descriptions, so that the correlation between symptom and disease was zero. The nurses nonetheless concluded that the correlation was positive and could, of course, point to many instances in the series to support this erroneous conclusion. Equally numerous instances of a lack of association between symptom and disease were forgotten or neglected. In Bayesian language, the error made by these clinicians was to focus excessively on the strength of association between the symptom and a disease, $P(S|D)$, when attention should properly be directed to that association in relation to the probability of observing that finding under other conditions as well. The diagnosticity of a cue depends on the likelihood ratio, $P(S|D_1)/P(S|D_2)$, not simply on the probability of a symptom given a particular condition. This is a difficult point to remember in informal clinical information processing and Bayes' theorem makes it explicit.

Excessive Data Collection

Clinical decision making depends on collecting relevant data, but sometimes it is not clear what is relevant. Clinical findings are often correlated with one another, since they are effects of an underlying common cause. In this circumstance, a number of cues are in effect redundant and can provide little additional information on logical or statistical grounds. The clinical decision could be made just as well by a formula that used fewer variables but weighted them properly. For example, one study of a battery of twelve laboratory tests [27] showed that most of the meaningful information could be accounted for by a formula for weighting and combining results on just four of these tests.

In data collection, there is a balance or tradeoff problem. Research has shown that when clinicians collect redundant cues or information with low validity, there is an increase in their confidence in their diagnostic judgments with little or no increment in diagnostic accuracy [20, 14]. From the clinician's viewpoint, however, there is a rationale for collecting redundant information. It is recognized that many cues have low reliabilities. For example, the test performance of children may be unstable due to transient shifts in motivation, attention and comprehension. The reliability of many cues in physical diagnosis is not terribly high [16]. Hence, it is reasonable to check a finding twice. On the other hand, if cues are stable and have high reliabilities, repeat observations are redundant and unnecessary. This line of argument leads to two important questions: How accurately do clinicians discriminate reliable from unreliable cues? How well calibrated are clinicians with respect to cue reliabilities? Empirical work is much needed.

Again, if the probability of a symptom given a disease [$P(S|D)$] is low, then the posterior probability of a disease given some findings can best be established by collecting multiple cues and updating opinion either sequentially or at the conclusion of data collection. On the other hand, if $P(S|D)$ is high, a single likelihood ratio may be close to sufficient for reaching a conclusion.

When the procedures for collecting redundant information are non-invasive and relatively inexpensive, such as taking a history or performing a physical examination, the decision maker may well be uninterested in a more efficient system for information processing, for the costs of developing and maintaining it may exceed foreseeable savings or gains. But if expensive or invasive procedures are involved, careful analysis of the inference process is warranted. So, for example, Neuhauser and Lewicki [18] have shown that the clinical practice of obtaining six stool guaiacs to test for colon cancer adds very little diagnostic certainty to that provided by two tests and at a substantial increase in cost.

Excessive numbers of tests may be ordered for another reason, limitations on the human capacity for indirect inference. A problem solver often prefers to seek direct evidence of what could be logically deduced from data already gathered, leading to collecting more data than would be needed by a more efficient information processor. This is yet one more consequence of the bounded rationality of the human information processing system.

Excessive data collection may impede the process of clinical inference by overloading the system's capacity. Accurate decision making depends upon both collecting and properly interpreting clinical data. The interpretive process can be adversely affected by collecting too much data, for the sheer volume of facts may impair the clinician's ability to sort out and focus upon the relevant variables. Decision trees, flow charts or algorithms can help to focus attention on the information that is truly relevant to the decision at hand [22].

Problems in Combining Information

Erroneous diagnostic or treatment decisions come about not only because mistakes are made in collecting data but also because of errors in interpreting single cues or combining several cues into a decision.

Some studies have shown that problems of integrating and combining information are a more important cause of diagnostic error than a simple lack of thoroughness in the workup of a patient [9]. Further, errors in information integration cannot necessarily be resolved by collecting more data; indeed, the more thorough the data collection, the less likely it is that all will be used in helping to structure the logical sequence of a clinical problem. Decision trees help to identify what data will be needed at what points in the problem and to insure that the clinician pays attention to these data at the right time and with proper weights. By conceptualizing changes in diagnostic opinion as changes in probability, the Bayesian approach focuses on collecting only that information that can lead to meaningful, cost-effective changes in opinion.

Conservatism and Liberalism

In studies of revising subjective probabilities, it has been repeatedly shown that human judgment has great difficulty matching the achievements of Bayes' theorem. Much doubt is thus cast on our capacities as intuitive statisticians [14]. When probabilistic data are processed simultaneously, conservatism in inference is generally observed. The impact the data should have on revising prior opinion is not adequately recognized and the opinion is updated less than is warranted according to Bayes' theorem [3, 23]. On the other hand, when data are processed sequentially, as is the case in most clinical reasoning, the experimental evidence suggests that there will be a tendency to overpredict, as uncertainty in the data is apparently ignored and each cue is treated as perfect or nearly perfect information [8]. These mistakes or biases in unaided human inference can be removed by any of a number of statistical procedures, notably by using Bayes' theorem or regression equations.

LIMITATION OF THE RESEARCH PARADIGM

Most of the psychological research on clinical judgment has been concerned with the question of how a set of data is or should be combined into a diagnostic conclusion or decision. The research commonly compares human performance to the decisions reached by statistical rules. The statistical approaches most commonly used are Bayes' theorem and regression equations derived from samples of judgments previously made. A great deal of this research has shown that statistical formulas can adequately model the meaningful variance in diagnostic judgment and can even surpass the accuracy of the judge from whose decisions a particular judgmental policy has been constructed [4, 10, 11, 17, 23]. So far, much has been said in favor of formal statistical approaches to clinical inference. It is time to balance the scales and give the clinical method its due.

In most of the research Scriven cites, the task of the clinician is to render a judgment, usually a diagnostic classification or a prediction, on the basis of a set of data provided by the experimenter. Twenty years ago, Holt [13] argued cogently that the ground rules of this research paradigm were constructed so as to favor the actuarial approach. The judgment task is defined as one of combining a set of data, a task that might well be accomplished mechanically. The research paradigm slights other crucial aspects of clinical activity, such as determining what data are needed and gathering them effectively. These tasks, Holt claimed then and clinicians continue to claim, are fundamental aspects of clinical activity and cannot be ignored in any relevant research.

This criticism should be remembered in evaluating the common research finding that the policies of clinicians can well be represented by linear models, despite the clinical intuition that the reasoning is highly configural and complex. From the point of view of the research, clinicians overestimate the complexity of their judgments. Why should this be so? In part, it may be because of the natural human tendency to magnify the importance and complexity of one's work. But we should recall that the research has focused mainly on combination rules. Clinicians must also be concerned with the complexities of data collection and with the problems of interaction with a patient or client while data are being collected. Research on clinical inference has to some extent simplified the task, as all experimental research simplifies reality. When the complexities of managing the human interaction, determining what data are to be collected in a limited time, and dealing with problems of value are added to the judgmental problem, the size of the task imposes such cognitive strain that it may seem to the clinician that a complex inference rule must be required. When a sequence of decisions is laid out, when values are analyzed separately from diagnostic states, or when attention is restricted to the problem of data integration, it becomes clear that the inference rule need not necessarily be complex.

Neither the decision-theoretic nor the policy-capturing approaches that have so dominated research on the psychology of clinical inference provide much guidance as to which diagnostic or treatment alternatives ought to be evaluated. They are more helpful in studying the process of data integration than in constructing a problem space. It appears that the paradigm most useful for descriptive and explanatory purposes does less in helping us to overcome the defects in the inference process identified earlier. Those models that offer the most direct advice about how to improve clinical inference, on the other hand, lay the weakest claim to reproducing the thought processes actually employed by clinicians in drawing inferences. The regression approach surely simplifies the clinical task and enthusiasm for its conclusions ought to be tempered by recognition of the focus and limitations of research. The information-processing and decision-theoretic approaches to the study of inference have identified other problematic issues. Awareness of the limitations on human rationality does imply that useful, perhaps even powerful, assistance to intuitive clinical inference can be provided by actuarial methods that incorporate frequency estimates or cue/disease correlations based on larger samples than would be possible for the average clinician to accumulate and without the distortions or simplifications of memory that are inevitably part of the repertory of clinical experience.

Michigan State University
East Lansing, Michigan

BIBLIOGRAPHY

1. Bartlett, F. C.: 1932, *Remembering: A Study in Experimental and Social Psychology*, Cambridge University Press, Cambridge.
2. De Groot, A. D.: 1965, *Thought and Choice in Chess*, Mouton, The Hague.
3. Edwards, W.: 1968, 'Conservatism in Human Information Processing', in B. Kleinmuntz (ed.), *Formal Representation of Human Judgment*, Wiley, New York, pp. 17–52.
4. Einhorn, H. J.: 1972, 'Expert Measurement and Mechanical Combination', *Organizational Behavior and Human Performance* **7**, 86–106.
5. Elstein, A. S.: 1976, 'Clinical Judgment: Psychological Research and Medical Practice', *Science* **194**, 696–700.
6. Elstein, A. S., Shulman, L. S., Sprafka, S.A.: 1978, *Medical Problem Solving: An Analysis of Clinical Reasoning*, Harvard University Press, Cambridge, Mass.
7. Fryback, D.: 1974, *Use of Radiologists' Subjective Probability Estimates in Medical Decision Making Problem*, Mathematical Psychology Program, Department of Psychology, University of Michigan, Ann Arbor.

8. Gettys, C. F., Kelly, C., and Peterson, C. R.: 1973, 'The Best-Guess Hypothesis in Multistage Inference', *Organizational Behavior and Human Performance* **10**, 364–373.
9. Gill, P. W., Leaper, D. J., Guillou, P. J., Staniland, J. R., Horrocks, J. C., and De Dombal, F. T.: 1973, 'Observer Variation in Clinical Diagnosis – a Computer-Aided Assessment of its Magnitude and Importance in 552 Patients with Abdominal Pain', *Methods of Information in Medicine* **12**, 108–113.
10. Goldberg, L. R.: 1968, 'Simple Models or Simple Processes? Some Research on Clinical Judgments', *American Psychologist* **23**, 483–496.
11. Goldberg, L. R.: 1970, 'Man Versus Model of Man: A Rationale, Plus Some Evidence, for a Method of Improving on Clinical Inferences', *Psychological Bulletin* **73**, 422–432.
12. Hammond, K., Stewart, T., Bremer, B., Steinmann, D.: 'Social Judgment Theory' in M. F. Kaplan and S. Schwartz, (eds.): 1975, *Human Judgment and Decision Processes*, Academic Press, New York, pp. 271–312.
13. Holt, R. R.: 1958, 'Clinical and Statistical Prediction: A Reformulation and Some New Data', *Journal of Abnormal and Social Psychology* **56**, 1–12.
14. Kahneman, D., and Tversky, A.: 1973, 'On the Psychology of Prediction', *Psychological Review* **80**, 237–251.
15. Kleinmuntz, B.: 1968, 'The Processing of Clinical Information by Man and Machine', in B. Kleinmuntz (ed.), *Formal Representation of Human Judgment*, Wiley, New York.
16. Koran, L. M.: 1965, 'The Reliability of Clinical Methods, Data and Judgments', *New England Journal of Medicine* **293**, 642–646.
17. Meehl, P. E.: 1954, *Clinical Versus Statistical Prediction*, University of Minnesota Press, Minneapolis.
18. Neuhauser, D., and Lewicki, A. M.: 1975, 'What do we Gain from the Sixth Stool Guaiac?', *New England Journal of Medicine* **293**, 226–228.
19. Newell, A., and Simon, H. A.: 1972, *Human Problem Solving*, Prentice-Hall, Englewood Cliffs, New Jersey.
20. Oskamp, S.: 1965, 'Overconfidence in Case Study Judgments', *Journal of Consulting Psychology* **29**, 261–265.
21. Simon, H. A.: 1969, *The Sciences of the Artificial*, MIT Press, Cambridge, Mass.
22. Sisson, J. D., Schoomaker, E. B., Ross, J. C.: 1976, 'Clinical Decision Analysis – the Hazard of Using Additional Data', *The Journal of the American Medical Association* **236**, 1259–1263.
23. Slovic, P., and Lichtenstein, S.: 1971, 'Comparison of Bayesian and Regression Approaches to the Study of Information Processing in Judgment', *Organizational Behavior and Human Performance* **6**, 649–744.
24. Slovic, P., Rorer, L. G., and Hoffman, P. J.: 1971, 'Analyzing Use of Diagnostic Signs', *Investigative Radiology* **6**, 18–26.
25. Smedslund, J.: 1963, 'The Concept of Correlation in Adults', *Scandinavian Journal of Psychology* **4**, 165–173.
26. Tversky, A., and Kahneman, D.: 1974, 'Judgment under Uncertainty: Heuristics and Biases', *Science* **185**, 1124–1131.
27. Zieve, L.: 1966, 'Misinterpretation and Abuse of Laboratory Tests by Clinicians', *Annals of the New York Academy of Science* **124**, 563–572.

ELLIOTT SOBER

THE ART AND SCIENCE OF CLINICAL JUDGMENT:
*An Informational Approach**

I

Is medical diagnosis an art or a science? This question alternatively appears puerile and formidable. On the one hand, the distinction between arts and sciences appears obvious, and pigeon-holing diagnosis a matter of course. Diagnosis looks to be more like physiology and psychology than like music, painting, or short story writing. If important questions are supposed to be hard and require considerable care in answering, our initial query does not seem to be one of the central subjects of the philosophy of medicine.

On the other hand, the distinction between art and science appears to have the earmarks of a difficult philosophical problem. Perhaps an analogy from the theory of knowledge would show why this is so. *Examples* of knowledge are not hard to find, and examples of the opposite condition are likewise fairly familiar. I would guess that as you now read this paper on the philosophy of medicine, you do not have the slightest inkling as to what the last word of the paper will be. However, generalizing from these and other examples has proved to be incredibly hard. A *theory* of knowledge is as evasive as examples of knowledge are commonplace. The distinction between art and science likewise has its obvious examples, but a general theory of what distinguishes them as kinds of activities is at best in its fledgling stage.

It is customary to divide medical diagnosis into its clinical and laboratory aspects and to relegate the former to art and the latter to science. A more finegrained taxonomy might then subdivide clinical and laboratory procedures into a number of techniques and characterize each as science or art. Such a classification might be forged by fiat, but to what point? Philosophical taxonomies, like those in biology or any other science, are useful to the degree that they *explain*, and to do this they should pick out fundamental differences. Without some systematic account of how art and science differ, answers to our question about medical diagnosis will be more or less intuitive but will not thereby prove to be very instructive.

When one turns to the writings of physicians on the question of whether clinical judgment is an art or a science, the problem of explaining the distinction shrinks to more modest proportions.[1] Usually the assertion that medicine

H. T. Engelhardt, Jr., S. F. Spicker and B. Towers (eds.), Clinical Judgment: A Critical Appraisal, 29–44.

is an art or that it is a science is used as a catch phrase for a specific thesis about the methods of reasoning employed. The interest of these assertions is less a matter of how medicine compares to music or mineralology than it is a matter of whether the claims pick out distinctive features of diagnosis. Sometimes when people say that diagnosis is a science, not an art, they mean to be saying that it is possible to describe precise quantitative techniques which generate differential diagnoses from specifications of symptoms and laboratory reports. Sometimes when people say that medicine is an art, not a science, they mean that it is a fundamentally qualitative and intuitive form of thought which is seriously distorted and impaired by the attempt to use precise mathematical models. The interest of the art versus science dispute in the philosophy of medicine mainly consists in the truth of such specific assertions as these. How our answers to these questions fit into some more general picture of the artistic and the scientific is of secondary importance.

In this paper, I propose to examine some of the particular versions that claims for and against the artistic or scientific character of clinical judgment have assumed. I shall try to divine what truth there might be in these assertions. Along the way, I hope to sketch a view of the nature of this debate that will bring out the real issue. If this thesis is correct, several consequences will follow for the conduct of research on clinical judgment and for the conception of the role of medical education.

Before beginning, a point of clarification. When one talks about the distinction of art and science, what is usually at stake is *not* a distinction between kinds of objects. Paintings may strike one as obviously being works of art, and equations as the very soul of science. But a painting can be subject to scientific scrutiny and a set of equations may be treated as wonderful poetry. What is at issue is not the kind of object but the kind of thinking people do about objects. We are interested in the habits of mind that medical people bring to bear on clinical problems and in understanding in what sense those techniques of thought are commonly claimed to be artful or scientific.

II

An extreme version of the diagnosis-is-an-art thesis is the following: The skilled clinician is capable of achieving an intuitive insight that is inherently non-logical. After the clinician figures out the correct causal explanation of the symptoms and laboratory data, he may be able to explain how he arrived at his judgment. But this *ex post facto* description of his act of mind is deficient in at least two ways. First, it may well be a rationalized reconstruction of a process whose actual characteristics are in no way logical or inferential. And second, the rationalized reconstruction, were it to be formulated into a general rule of medical evidence, would prove to be extremely unreliable when applied in a great number of cases. This is merely to say that the clinician's description of what it is like for him to make clinical diagnoses is a rationalization. It is false to the facts of his own psychological processes and would moreover be bad medicine if generally incorporated into one's corpus of clinical procedures.

A stunning and unexpected diagnostic solution to a problem is not, of course, the most frequent clinical event in a physician's everyday practice, but it happens often enough for everyone to be familiar with it. The occurrence of creative insights of this kind quickly raises the question "How did he do it?", and the physician's own conjectures about his own thought processes will often fail to satisfy. In the light of this ignorance, it is tempting to claim: There is no such thing as a *method of reasoning* at work. In mundane, "easy" clinical settings, one's thought processes may in fact obey some recipe of rational calculation. But in exceptional circumstances where exceptional skill is called for, no method of obtaining an answer exists, and so the physician produces his hypothesis in some other way. And how might he do this? By an intuitive creative insight.

Philosophers of science have been tempted to say this kind of thing when confronted by the acts of mind that seem to be involved in revolutionary scientific insights. How did Einstein arrive at his special theory of relativity? His descriptions of his own methods of thinking seem to be radically incomplete. If we were to endow another scientist with what Einstein described as his way of proceeding, we would not expect this other fellow to come up with a suitably impressive scientific insight. And if asked why, we would say – because he is no Einstein. This perhaps points to our tacit sympathies with the view that creativity knows no method. Einstein succeeded not in virtue of following any recipe but in virtue of being the kind of thinker he was. Although recipes may perhaps exist for the more modest and mundane areas of scientific inquiry, there is no logic of scientific discovery. Scientific creativity, like exceptional clinical insight, proceeds by breaking, not following, rules.[2]

Let me say flat out that I think that this position is a dodge. If there is any truth at all in the claim that there is no recipe for finding accurate diagnoses or explanatory theories, it is limited to the rather more modest assertion that we do not know what general methods of problem solving are involved in creative thinking. But let us consider the various alternative hypotheses that might compete with the claim that creative insights are produced by the application of general inferential techniques. First there is the null hypothesis, which asserts that creative diagnoses are not the upshot of any causal process whatsoever. Second there is the family of hypotheses which assert that some causal mechanism is responsible for the emergence of insightful diagnoses, but that this method somehow bypasses the cognitive systems engaged in problem solving and hypothesis formation.

The null hypothesis is presumably the hypothesis of last resort. It would make a miracle of whatever uniformities there appear to be in the diagnoses clinicians produce given symptomatic and laboratory evidence. The noncognitive hypothesis, though not itself a concession of defeat, is implausible for another reason. Normal, routine problem solving, in its clinical, scientific, and everyday form, occurs within the context of the human information-processing system. The noncognitive hypothesis sketched above would have it that this system suddenly ceases and desists when a truly formidable extraordinary problem presents itself. From the design standpoint, it is unlikely that we are composed of different processing systems, one that gets used when the questions are easy, another when they are hard.

What then are we to make of the introspective reports that clinicians or scientists produce as descriptions of the thought processes that lead them to their insights? These descriptions are to be accorded the same epistemological status that we attribute to their hypotheses about the patient or about the phenomenon to be explained. These introspective reports issue from the information-processing system's turning on itself, generating representations of what it itself has done. These representations may be as distorted, inaccurate, and unilluminating as any hypothesis. But we should not conclude that inference and reason go on holiday when creative insight is called for.

The general picture I am invoking here will strike some as obvious and others as barely intelligible. Clinical judgment is to be understood as occurring within an information-processing system, which has as its input a specification of observed characteristics of the patient and perhaps some laboratory data, and has as output a differential diagnosis. This is a fundamentally inductive process in which the differential diagnosis is a list of hypotheses each of which would explain the symptoms and laboratory data. This inferential picture of the diagnostic process should be viewed as both a logical and a psychological thesis. As logic, it describes what it would be rational for physicians to do. As psychology, it describes the intervening mechanisms that account for how diagnostic hypotheses are arrived at on the basis of other data. This is not to say that physicians are perfectly rational in the way they process information. What I am claiming is that the kinds of models appropriate to both description and prescription are fundamentally the same.

Another claim one often hears in defense of the view that clinical judgment is an art not a science invokes the idea that each patient is unique and goes on to assert that scientific generalizations about what all individuals of a certain kind are like must fail to make contact with the nexus of the diagnostic problem. What are we to make of this idea that clinicians work on the particular while scientists focus their gaze on the general? First, it should be noted that some sciences, astronomy for example, have particular individuals as their objects of inquiry. And second, applied science, engineering for example, has never felt that constructing a single bridge presents any conceptual difficulties for the application of general knowledge. Science, applied in practice, does not cease to be science, nor do the methods of making such application seem to involve any fundamental departure from scientific conceptions of hypothesis and confirmation.

The parallel with engineering is instructive. The *kind* of object dealt with in applied science, medicine included, is an individual, whether it be bridge or patient. In repairing a bridge one makes use of the results of scientific theories; one formulates an explanation of what has gone wrong; one musters technique to provide a remedy. In this respect, medicine and engineering differ from biology and physics. In these latter, so-called "pure" endeavors, the object of inquiry is dramatically different. Here one hopes to arrive at empirical generalizations and laws of nature which are characteristically *general* in scope. Rather than arriving at hypotheses about *this* bridge or *this* patient, the scientist in search of natural regularities will try to find out what *all* objects of a certain kind are like. This distinction between areas of inquiry in which an individual constitutes the object of conjecture and areas of inquiry in which generalizations about kinds of objects are the focus of research should not be taken to imply that there are special *logics* at work in these different disciplines. Deduction, induction, and inference to the best explanation (what C. S. Peirce called "abduction") play a role in *all* of science; more on this later.

But let us go back to the concept of uniqueness. What does it mean to say that a patient is unique? If one is merely saying that the patient has a property that no other individual has, uniqueness can be granted without further ado. *Every* object is unique in some respect or other. And just as truistically, we can say that no object is unique in all respects. But such facile flippancy aside, what idea are people attempting to express by appeals to uniqueness? Possibly it is something like this: Let us conceive of diagnosis as a problem-solving situation in which the set of all possible answers is the set of all known diseases. The problem is to decide which of these diseases the patient has. We now want to know how much information about the patient is required before a diagnosis can be rationally determined. The more unique the patient is, the greater the amount of information required before rendering a diagnosis.[3] The limiting case of uniqueness would make diagnosis, and indeed all inductive extrapolation, impossible. Maximal uniqueness obtains when *all* the information about a patient is required in advance of a diagnostic opinion; were this complete characterization ever achieved, there would be no diagnosis left to be made. Presumably, only an inductive sceptic would be much attracted by the claim that the patient is maximally unique. However, notice that any degree of uniqueness less than this is perfectly consistent with seeing clinical judgment as a scientific activity. Within any science, answering the question "Which of a group of properties does this individual have?" will inevitably require extra information about the individual, and the individual considered, whether it be animal, vegetable, or mineral, will thereby partake of some degree of uniqueness.

There is another aspect of this issue of uniqueness that needs to be noted. We have construed uniqueness in terms of how much information about an individual is required before it is rational to hazard a hypothesis. This question is affected by two considerations. On the one hand there is the question of how good the indicators are with respect to your question. If one wants to know which of a family of properties an individual has, how much information one needs before venturing an answer will depend in part on how much information the indicators bear with respect to your question. It may be that the indicators are probabilistic in such a way that a great many details are required before the probabilities of different answers are enough distinguished. A second element in this uniqueness phenomenon is ethical. Sometimes how much information is needed before an answer can rationally be given will depend on the ethical consequences of being wrong. If our choice from a set of alternative answers to our diagnostic question will have immense consequences for the life and well-being of the patient, then it behooves us to require considerable information before reaching an answer (that is, if time permits). And to the degree that the alternatives are very close in their associated utilities, less in the way of corroborating information is essential.[4]

We have distinguished between two sources of uniqueness, which we might label indicator uniqueness and ethical uniqueness. We argued earlier that indicator uniqueness is perfectly consistent with clinical diagnosis being a science. I think the same can be said for ethical uniqueness. What follows from uniqueness phenomena due to ethical issues is that ethical considerations must be part of one's rational calculations about how much evidence is enough. But this no more removes diagnosis from the domain of science that engineering is rendered unscientific by the fact that in building bridges one has to worry about what would happen if one were mistaken about calculations of stress.

A third source of the idea that diagnosis is an art not a science focuses on the role of the emotions. It is not uncommon for people to think of reason and emotion as distinct, the former capable in principle of acting without the latter. That is, although the bloodless scientist is a myth, it is sometimes thought that if a person could somehow be parted from his emotional reactions, he could nevertheless succeed in doing scientific research. It is sometimes even held that his research record would be improved. This, of course, is closely akin to the idea discussed above that scientists need make no value judgments. And just as I above argued that scientific diagnosis must assign a place to assessing the ethical consequences of error, so now I want to suggest that clinical diagnosis cannot proceed without the emotions playing their part.

It would seem that empathy with electrons is inessential for research in particle physics. A scientist with no such attachment would perhaps be lacking in romance, but his insights need not suffer. However, when the domain of inquiry is people and their diseases, and the diagnostician a person himself, the role of the emotions is transformed. In diagnosis, the emotions are a source of information; they function cognitively. A person somehow cut off from emotions would be deprived of data, and this might have the same kind of debilitating effect on diagnosis as the absence of a laboratory test. It is misleading, I think, to stress the importance of sensitivity by saying that diagnosis is not just a science. Rather, one should explain the importance of the emotions by placing them within the context of the scientist's need to gather relevant information by the best means he has available.[5]

Another much touted dichotomy that often lies behind people's distinction of art from science is the qualitative versus the quantitative. The clinician is engaging in the art of healing when he accumulates evidence and makes inferences that involve purely qualitative characterizations of symptoms and postulated disease. When he turns to mathematics, either by appealing to the quantitative analysis of a laboratory test or by applying statistical information about the actuarial profile of a given population, he is acting as a scientist. This distinguishing of aspects in the clinician's work is often accompanied by some view of which is better. If it is qualitative techniques that one wants t laud, we pay them compliments by saying that they avoid needless precision and are more in touch with the patient. When quantitative techniques are to be the object of praise, we say they are scientific, verifiable, and relevantly precise. But testimonials aside, what distinguishes the quantitative from the qualitative? A logician aware of the intimate connections that have been discovered in the last 100 years between logic and mathematics cannot help being puzzled by the considerable prestige this dichotomy seems to enjoy. Russell, Whitehead, and Frege tried to show that the whole of mathematics is nothing but logic; they did not meet with complete success in this enterprise, but certain intimate connections between logic and mathematics are now beyond doubt. If we ignore some of the difficulties that the attempt to reduce mathematics to logic has encountered, the following appears to be uncontroversial: Inference using purely qualitative concepts can be just as precise as the most finely honed mathematics. Moreover, the modes of reasoning that are used in mathematics are also used in the most mundane syllogisms. Mathematics has no special logic.

Given the approximate truth of Russell's thought that there is a seamless web between logic and mathematics and the more familiar fact that quantitative description can be accurate and quantitative characterization obscurantist, one wonders what difference between the quantitative and the qualitative could have excited so much interest. No small measure of the allure of quantitative methods is to be chalked up to physics-worship. Physics is a science, physics is mathematical, therefore science is mathematical. We may pass over this syllogism without comment. Nevertheless, it is a striking fact about the history of science that maturing of theory is often accompanied by the replacement of qualitative by quantitative concepts. There are exceptions to this rule of thumb which should be borne in mind, like Chomsky's innovations in linguistics. But even so, the utility of quantitative methods is impressive. What does the shift from qualitative to quantitative consist in?

Quantitative statements are capable of conveying more information than are qualitative ones. A physical magnitude like temperature has as many different values as there are real numbers, and that is why saying what an object's temperature is conveys so much information: In saying that the temperature is 98° F., you exclude all values different from 98°, and that is to exclude an awful lot. Roughly, the more possibilities your statement excludes the more informative it is.[6] That is why saying that a patient either has pneumonia or does not is uninformative. A family of qualitative properties, colors for instance, has much fewer alternatives, and so attribution of color to an object carries less information.

In saying that the information-carrying capacities of qualitative and quantitative descriptions differ in this way, I am not saying that every quantitative description is more informative than any qualitative one. Saying that an object is green is no less informative than saying that its characteristic wave length is between this value and that. And to say that the probability of rain is between 0 and 98% is to say practically nothing at all. Quantitative concepts give one the opportunity of being precise, but this opportunity need not be exploited in every instance.

Another way in which quantitative statements can provide little information is by providing needless precision. In no area of science, medicine included, is one interested in being as accurate and informative as possible, full stop. Rather, one wants to solve certain problems, and precision which is relevant to their solution is much prized. In using quantitative techniques, one must guard against making very informative statements that provide virtually no information relevant to answering one's questions.

We have explored four pairs of dichotomies that often lurk in the back of people's minds when they think about clinical judgment as an artistic or scientific endeavor. The idea that clinical judgment involves some kind of nonlogical intuitive insight has been deflated by seeing that it has no more evidence on its side than the fact that we are now very much in the dark about how human beings solve problems. But ignorance of the logic of clinical discovery is no reason to think that there is no such logic. With respect to the other alleged dichotomies, we have tried to characterize the uniqueness of the individual, the role of the emotions, and the use of qualitative concepts in such a way that clinical judgment can be seen as a scientific undertaking.

Sometimes a scientific procedure is understood to be one in which a purely mechanical procedure is known to do the job; there is a science of long division, but clinical judgment is an art. If scientific techniques are limited to the mindlessly computational, then very little counts as science, including the activity of science itself. In discussing the artful and the scientific aspects of diagnosis, one has to be careful not to short-circuit the discussion by characterizing the dichotomy in a wholly unrealistic way. The attempts described so far have not cast very much light on our problem. I do not think that the widespread discussion of this question is to be chalked up to philosophical confusion. Clinicians *are* picking up on something when they talk about the art and science of diagnosis. But what is the underlying reality that forms the object of this debate?

III

The kind of model I have been invoking as a description of clinical diagnosis is familiar within psychology. Information-processing models are much used in the study of visual perception, language learning, problem solving, and memory. In all such models, a given cognitive capacity is understood in terms of a computational process in which representations are formed and transformed. In visual perception, for example, one might try to show how knowledge of the three-dimensional world of objects is obtained by inference from the data provided by a series of two-dimensional retinal patterns. In language acquisition, one might try to show how a child's knowledge of the grammar of his language is inferred from the sample of sentences he actually hears. To use this model as a description of what occurs in clinical diagnosis, one would try to describe diagnostic hypotheses as inferred from the data obtained in clinical examination and from the reports of laboratory tests. In some instances, physicians are themselves conscious of proceeding inferentially; in other cases, the occurrence and nature of the inferences made are somewhat opaque to introspection. In both cases, we might conjecture, the psychological processes are fundamentally the same.

There are two broad kinds of justifications that can now be given for the use of information-processing models in these various domains of cognition. On the one hand, the methodological objections occasionally raised against such models can be refuted. Behaviorists sometimes criticize the mentalistic postulates of these models by alleging that they are empty or unverifiable; materialists sometimes object that the only legitimate models are ones which deal explicitly in the chemical nature of the brain. Both of these criticisms can be met, I think; I will not go into how this is to be done. The second kind of justification of information-processing models is more positive. Rather than refuting somewhat *a priori* philosophical objections, this technique of argument points to the fruitfulness and success that such models have enjoyed and goes on to ask: What other kinds of models can claim to have achieved so much or promise to prove fruitful in the future? Behaviorist and materialist objections to our mentalist approach would carry more weight if they could point to developing and fruitful research programs of their own. But this is not the case in areas like problem solving and language acquisition. So, information-processing models are immune from *a priori* philosophical objection. They have proved to be and promise to be fruitful. And they are the only currently developed approaches that seem to have a chance. This rough sketch of the kinds of justification that can be invoked does not guarantee that such models will certainly prove correct. But they are now the best bets.

Given the legitimacy of this kind of approach, one very fundamental question that arises is how the processors used in different cognitive domains are related to each other. In visual perception, perceptual judgments about physical objects are inferred from retinal information; in problem solving, answers to questions are inferred from given starting assumptions. Are the techniques of inference whereby information is represented and transformed the same across these different cognitive capacities, or do they all share in the use of certain general problem solving techniques? It is premature to try to answer this question with any amount of certainty, although this has not stopped Chomsky, for example, from asserting that the modes of inference used in language acquisition are distinct from the kinds of problem-solving strategies used in other areas of cognition.[8] Without trying to definitively settle this rather complex question, let me try to suggest one kind of evidence that is relevant and to relate this idea to a way one might try to understand what physicians do when they process information and come up with diagnoses.

If the kinds of errors that people make in a given kind of problem-solving situation are very much like the kinds of errors that clinicians are prone to, then we have some evidence for the claim that the underlying mechanisms that are at work in everyday problem-solving are very much like the kinds of mechanisms used in clinical diagnosis. Let me conjecture that this close resemblance exists. I thereby deny that there is any such thing as a special logic of medical diagnosis. As a matter of psychology, clinicians make use of standard inferential equipment which is subject to the same kinds of imperfections and breakdowns that standard equipment encounters in other domains of thought.

If this is right, then we should be able to discover what kinds of errors and imperfections the diagnostic process is subject to by looking at the psychological literature on problem solving. Let me give an example of how this transferral of results from psychology in general to the area of clinical diagnosis might proceed. In their pioneering work *A Study of Thinking*, Bruner, Goodnow, and Austin [2] discovered that human beings find it very difficult to make use of negative evidence and to use disjunctive hypotheses. They pointed out, and more recent studies ([3, 21, 22]) have tended to substantiate the claim, that when individuals are faced with a pair of alternative hypotheses X and Y, there is a very strong tendency to try to narrow the field to a single hypothesis by attempting to confirm X or to confirm Y. A less automatic strategy is to try to attain a unique hypothesis by *dis*confirming *X* or *dis*confirming *Y*. This indirect process is often avoided, even when the direct strategy is more cumbersome and less efficient.

Let me describe an experimental paradigm that has been used [21] to support this general claim. The subject is shown a triplet of numbers, say 2, 4, 6, and is told that these numbers obey a rule. The subject's task is to discover the general rule by producing triplets of numbers of his own devising and asking the experimenter whether they conform to the rule. The subject is to state what he thinks the rule is only after he has accumulated what he judges to be enough data. Here is a simple protocol which illustrates the tendency to ignore negative tests:

> It looks like the series 2, 4, 6 is generated by adding 2 to the first member of the series to produce the second, and by doing the same thing to the second member to produce the third. So let us try 3, 5, 7. Ah ha! The experimenter says that conforms to the rule too! Just to be safe let's try 11, 13, 15. Well that works also, so it looks like the rule is to add 2 twice over to any initial number,

The subjects in the experiment were more circumspect than this, but there was a substantial tendency to think in ways which this protocol embodies to an extreme degree.[9]

It has been argued [5] that these frailties of reasoning can affect clinical judgment. Clinicians sometimes fail to extract all of the relevant information provided by laboratory data and obtained from examining patients. This can lead to ordering unnecessary laboratory tests. When several hypotheses are in the field concerning the etiology of a set of clinical symptoms and laboratory findings, there is a tendency to try to confirm one hypothesis directly; the strategy of disconfirming alternatives is less automatic. If this is true, then we have some confirmation of our initial conjecture that clinical diagnosis involves the use of general problem solving techniques; there is no special logic of diagnosis.

This general working hypothesis points to numerous questions whose answers might illuminate and perhaps improve the clinical process. Let me give another example. Every clinician has at one time or another been in a situation in which the diagnostic answer evaded him, although in retrospect the answer was as plain as day. As hypotheses are developed we tend to lose our flexibility in inventing alternatives, so that even if the resulting hypothesis is known to be incorrect it is very hard to look at the data in a new way. Often this familiar dilemma can be remedied by bringing in someone who can provide a fresh perspective; hence the utility of consultation. When several physicians work together on a case, they usually share their conjectures and jointly develop a differential diagnosis. From the point of view of problem-solving strategy, this has advantages and disadvantages. The advantage is that in pooling knowledge and hunches, one can reach a collective decision more quickly than if each person develops his hypotheses separately and then the group pools and debates the separate results. However, the advantage of being able to more quickly reach a consensus has its price. In the process of working together from the beginning, alternative views are more quickly dismissed. A half-baked idea, challenged at conception, will be discarded, whereas if it were pondered by its adherent for a little longer, it might turn into something significant. The trade-offs between these different problem-solving strategies are fairly apparent. It would be interesting to see some sort of general argument which would allow one to decide in what kinds of circumstances one is more productive than the other.

I have only been able to touch on a few of the details that need to be examined within the general program of inquiry I am advocating. As the last example makes clear, we need to look at the social context of collective decision making as well as the psychology of reasoning of individual clinicians. General patterns which emerge from examining medical thinking need to be compared with ideas in the psychology of problem solving and with logical models of how rational inference best proceeds. One of the points of pursuing these kinds of questions is to improve clinical practice, and one way in which this might be achieved is through our conception of how medical education should be organized. In closing, I want to make a single point about what our information processing model does *not* show about medical education; this will provide a tentative answer to our initial question: In what sense, if any, can medical diagnosis be said to be an art?

IV

If good diagnostic practice consists in the accumulation and inferential manipulation of information in conformity with general problem solving techniques, does it follow that the best way to learn how to be a skilled clinician is to study some explicit and detailed specification of these rules? Emphatically not. This strategy of learning may be just as cumbersome and inefficient as it would be to learn a language by studying an accurate and complete representation of its grammar. What does seem to be fruitful is to study and absorb certain fragments and general properties of inferential strategy. It is for good reason that medical textbooks provide only an elementary sketch of syllogistic inference. Anyone familiar with the complexities that must lurk in any approximately realistic nondeductive logic may perhaps wonder why things look so wonderfully simple in medical texts. The answer is that such texts only describe the tip of the iceberg, and rightly so. Even if a correct formulation of rational inductive policy were available, which it decidedly is not, it would be so complicated and intricate as to be fairly useless as a recipe for clinical practice.

This phenomenon is familiar in artificial intelligence research on chess playing programs. There are now in existence programs for chess playing that enable computers to be just as skillful as middle-level competitors [9]. No computer has the stature of a grand master, but some computers are pretty damned good.[10] In spite of this, I can assure you that your desire to improve your chess game would not be efficiently satisfied by my laying out the computer program and saying: Play like this. Rather, what one should do is play chess with a skilled player who is also a good teacher. The totality of what such a teacher asserts will fall short of the completeness and precision of the program he perhaps follows, but it will be information that you can make use of.

Let us now go back to the question of the art and science of diagnosis. Is skillful chess playing an art or a science? Presumably, for any degree of skill in chess, there is a computer program which can achieve that level of skill. Has chess suddenly turned into a mindless routine, like long division? Certainly not. We call chess an art because though chess playing recipes may exist, it is not within our information-processing capacities to play chess by absorbing and explicitly appealing to the dictates of such a program. Performing accurate long divisions, on the other hand, is no art at all, because the mechanical recipe for success is easily learned and efficiently applied. To defend the artfulness of clinical diagnosis against the charge that it is characterizable as a mechanical process, it is not necessary to assert that diagnosis is fundamentally alogical. This defense, as we have suggested, is a mystification. Rather, the artfulness of an activity turns on how the computational processes that people employ may be made the objects of self-conscious learning. Where the logic is sufficiently complex, explicitly learning and using the recipe is well-nigh impossible, even though the methodology does exist. In such domains, an indirect assault on the object of inquiry is more efficient. In learning how to speak a language, learning how to play chess, and learning how to make clinical diagnoses, one absorbs the theory indirectly by participating in the practice on which the theory is based. It is this fact about clinical diagnosis, and not its allegedly nonlogical, nongeneral, emotional, or qualitative characteristics, that makes it an art.

University of Wisconsin
Madison, Wisconsin

NOTES

* Murray Katcher, Lewis Leavitt, Ernan McMullin, and Daniel Wikler helped improve the formulations in this paper by making a number of thoughtful suggestions.

[1] Feinstein [6] is wary of reviving a dead horse only to flog it to death, but plunges into the art versus science dispute. His way of distinguishing artful from scientific components of clinical judgment does not appear to be very systematic.

[2] This idea is standard fare in 20th century philosophy of science. See, for example, Popper [14], Reichenbach [15], and Hempel [11]. For criticisms of the claim that there is no logic of discovery, see Hanson [10] and Sober [19]. For Einstein's description of how he did it, see Wertheimer [23].

[3] In Sober [20], I develop a notion of information according to which the more informative a hypothesis is in answering a question, the less extra information it requires to imply an answer. This notion of information can be used to give a more precise characterization of the idea of uniqueness sketched above.

[4] Rudner [16] defends the claim that the ethical consequences of error are relevant to deciding how much evidence is enough evidence to make acceptance rational.

[5] Goodman [8] argues that the emotions function cognitively in the way in which we come to understand works of art.

[6] This connection between the information content of a hypothesis and the number of possibilities it excludes was first made by Popper [14]. It is also found in the notion of semantic information developed by Carnap and Bar-Hillel [1], and in the nonprobabilistic measure of informativeness I develop in Sober [20].

[7] I discuss some features of information processing models in visual perception and language acquisition in Sober [18, 19, 20]. See also Fodor [7] for a philosophical and empirical defense of such models.

[8] See Chomsky [4] for discussion, and Sober [20] for criticism.

[9] Popper [14] has railed against this kind of fallacy by arguing that the scientific method is a logic of falsification, not verification. Although I have a number of reservations about his positive view, there can be no doubt that the kind of 'confirmation' described above is hardly confirmation at all. At the risk of belaboring the obvious, let me point out what has gone wrong in the sample protocol. To be sure, the rule arrived at conforms to the evidence available. The problem is that so do indefinitely many other rules. "Pick a triplet of numbers such that each is larger than the preceding" and "Pick three numbers less than 100" likewise fit the data. The fallacy of the protocol is that it never tried to *dis*conform the favored hypothesis, but only attempted to accumulate favorable evidence.

[10] According to Simon ([17], p. 63), the Greenblatt program [9] 'plays a respectable game in weekend tournaments and has a Class C American Chess Federation rating.'

BIBLIOGRAPHY

1. Bar-Hillel, Y. and Carnap, R.: 1964, 'An Outline of a Theory of Semantic Information', in Y. Bar-Hillel (ed.), *Language and Information*, Addison-Wesley, Reading, Mass., pp. 221–75.
2. Bruner, J., Goodnow, J., Austin, G.: 1956, *A Study of Thinking*, Wiley, New York.
3. Campbell, A.: 1965, 'On the Solving of Code Items Demanding the Use of Indirect Procedures', *British Journal of Psychology* **56**, pp. 45–51.
4. Chomsky, N.: 1965, *Aspects of the Theory of Syntax*, M.I.T. Press, Cambridge, Mass.
5. Elstein, A.: 1976, 'Clinical Judgment: Psychological Research and Medical Practice', *Science* **194**, 696–700.
6. Feinstein, A.: 1967, *Clinical Judgment*, Williams & Wilkins, Baltimore.
7. Fodor, J.: 1976, *The Language of Thought*, Thomas Crowell, New York.
8. Goodman, N.: 1968, *Languages of Art*, Bobbs-Merrill, Indianapolis.
9. Greenblatt, R., Eastlake, D., and Crocker, S.: 1967, 'The Greenblatt Chess Program', *Proceedings of the Fall Joint Computer Conference*, Anaheim, Cal., pp. 801–10.
10. Hanson, N.: 1970, 'Is There a Logic of Discovery' in B. Brody (ed.) *Readings in the Philosophy of Science*, Prentice-Hall, Englewood Cliffs, N.J., pp. 620–33.
11. Hempel, C.: 1965, *Philosophy of Natural Science*, Prentice-Hall, Englewood Cliffs, N.J.
12. Johnson-Abercrombie, M.: 1960, *The Anatomy of Judgement*, Penguin, Baltimore.
13. Murphy, E.: 1976, *The Logic of Medicine*, Johns Hopkins University Press, Baltimore.
14. Popper, K.: 1959, *The Logic of Scientific Discovery*, Hutchinson, London.
15. Reichenbach, H.: 1938, *Experience and Prediction*, University of Chicago Press, Chicago.
16. Rudner, R.: 1970, 'The Scientist Qua Scientist Makes Value Judgments', in B. Brody (ed.) *Readings in Philosophy of Science*, Prentice-Hall, Englewood Cliffs, N.J., pp. 540–6.
17. Simon, H.: 1969, *The Sciences of the Artificial*, M.I.T. Press, Cambridge, Mass.
18. Sober, E.: 1976, 'Mental Representations', *Synthese* **33**, pp. 101–48.
19. Sober, E.: 1978, 'Psychologism', *Journal for the Theory of Social Behavior* 8, pp. 165–193.
20. Sober, E.: 1975, *Simplicity*, Oxford University Press, Oxford.
21. Wason, P.: 1960, 'On the Failure to Eliminate Hypotheses in a Conceptual Task,' *Quarterly Journal of Experimental Psychology* **12**, pp. 129–40.
22. Wason, P., Johnson-Laird, P.: 1972, *Psychology of Reasoning: Structure and Content*, B. T. Batsford, London.
23. Wertheimer, M.: 1961, *Productive Thinking*, Tavistock, London.

D. L. ROSENHAN

WHEN DOES A DIAGNOSIS BECOME A CLINICAL JUDGMENT?

Verifiability, Reliability and Umbrella Effects in Diagnosis

To ask whether diagnosis is an art or a science is to make some interesting prior assumptions about the properties of diagnosis: That it is verifiable; that diagnosticians presented with identical data will concur in identical diagnoses; that the diagnosis usefully conveys information about the present status and prognosis of the patient; that it has treatment implications; and that it is not likely to be clouded by negative umbrella effects.

In this paper, I discuss some of these issues, using the current status of psychiatric diagnosis as my text. In the course of examining these matters, I hope to make convincing distinctions between three kinds of diagnosis: *Functional diagnoses*, which are the kind we make when we have treatment objectives in mind; *clinical diagnoses*, which are made with classification and communication as our intent; and *research diagnoses*, which are diagnoses that facilitate the research enterprise but have less than optimal bearing on the clinical one.

VERIFIABLE AND FUNCTIONAL CLINICAL DIAGNOSES

For those who rely primarily on quantitative indices for their diagnoses, and even for those who accommodate easily to qualitative data, nothing could be remote as primary evidence for a diagnosis than the data that emerge from those chimera that we call dreams. Nevertheless, given a skilled diagnostician and/or easily translatable symbols, even the materials of dreams can contribute usefully to a functional diagnosis.

Three instances come to mind, instances that illustrate the consummate skill of the clinician, or, alternately, the transparency of the symbol, and finally the utility of the diagnosis. The first dream is provided by the late Theodore Reik, the second by Paul Meehl, and the third from my own notes.

> One session took the following course. After a few sentences about the uneventful day, the patient fell into a long silence. She assured me that nothing was in her thoughts. Silence from me. After many minutes she complained about a toothache. She told me that she had been to the dentist yesterday. He had given her an injection and then had pulled a wisdom tooth. The spot was hurting again. New and longer silence. She pointed to my bookcase in the corner and said, "There's a book standing on its head." Without

H. T. Engelhardt, Jr., S. F. Spicker and B. Towers (eds.), Clinical Judgment: A Critical Appraisal, 45–56.

the slightest hesitation and in a reproachful voice I said, "But why did you not tell me that you had had an abortion?" ([10], p. 51)

The patient was amazed, and though disturbed that he "knew" so much, confirmed his intuition. And unless Theodore Reik knew a great deal more about this patient than he has told us – and memory being what it is, that is not the remotest possibility – there is no way in which we can presently explain such a diagnosis. The rules by which such a transformation might occur are simply unknown to us. What it is that idiosyncratically regulates symbol and metaphor has yet to be made clear. Yet, if we take him at his word, Reik is nonetheless able to use those symbolic and metaphoric data to arrive at a diagnosis. We call that "art". We might just as well call that "genius". Both terms are intended as compliments. And while both describe a true mystery, they should not be taken to imply mysticism, as Sober points out.

Meehl gives some sense of the transformation rules that might be involved.

A patient tells a dream which begins as follows: "I was in the basement of my parents' house, back home. It seems that I was ironing, and a fellow whom I never went out with, and hardly knew, had brought some shirts over for me to iron for him. I felt vaguely resentful about this – oh, and by the way, he was dressed in a riding habit, of all things" (grinning). Now, this patient had said in the preceding interview that it would be too easy to get into the *habit* of having sexual relationships with her present boy friend, and that since she did not really care a great deal about him, she must try to avoid this. If the phrase "riding habit" is a sexual pun, we infer that the adolescent acquaintance whom she "hardly knew" represents her present friend in the dream. The remainder of the dream and her associations to it, confirmed this hypothesis. ([10], p. 71)

The dream, we suspect, says more than Meehl indicates. Surely, ironing shirts for someone one hardly knows or likes must be pregnant with meaning (pun intended). Combined with the habit of riding, one senses not only what the dream means, but what the transformations might be.

Finally, a dream in which the transformations are quite apparent:

A woman is in love with, and loved by, two men, Jules and Michael. After much struggle, she decides to marry Michael. That night she has the following dream. "I dreamt that I was climbing a fire escape into my dormitory. It was raining. I was carrying a box under my raincoat, and hoping no one would see me. I got to my floor, opened the fire door, and tiptoed down the hallway to my room. I locked the room and finally took the box out from under my coat. I opened it up, and it was full of diamonds and rubies."

Where the transformation is obvious, that matter becomes less a clinical art. The data rise to meet the diagnosis, as it were. It takes no great intuition or art to understand the transformation in this dream, from Jules to Jewels. While the nature of the transforming mechanism is surely unclear, the transformation itself is patent. No mysticism here.

Two things are evident from these examples, then. Without going so far as indicating, with Sober, that we understand the cognitive processes here, except in the loosest sense, it is clear that there are transformations, and that these transformations are relatively difficult or easy to comprehend. That dimension of ease and difficulty defines whether the understanding will be seen as art in the positive sense.

But the examples, it seems to me, go further. They lay to rest the notion that useful psychiatric diagnoses of the clinical sort – functional and verifiable ones – can never be made. Indeed, the examples from dreams are sufficiently dramatic and compelling to make that notion seem a straw man. So be it.

FUNCTIONAL AND CATEGORICAL DIAGNOSES

Observe that the term diagnosis is used in both its formal and loose sense. The diagnoses that arise from these dreams allow a therapist to distinguish or discern one set of concerns that a patient has from the infinity of possible others, and to do so with some possibility of verification. Reik's patient is concerned with abortion and trust; Meehl's with meaningless sex and the possibilities of being abused; my own, with post-decision regrets. Such diagnoses are verifiable. And they are, moreover, useful in the sense that they provide guideposts to subsequent treatment.

From a different point of view, my use of the term is quite loose, for in no sense do these diagnoses lend themselves to formal categorical differentiations or classifications of the sort that are commonly used. Regarding such diagnoses, particularly in the area of psychiatry, I have been and am greatly ambivalent. And because I suspect that I am not alone in this, I elaborate on these issues briefly.

THREE KINDS OF DIAGNOSES

Three kinds of diagnoses can be discerned. The first is a functional diagnosis, and it is the most important. Such a diagnosis not only communicates what the problem is, but tacitly or openly suggests the remedy. The diagnoses that derive from the dreams fit this category.

A second kind of diagnosis shares properties with the first. It, too, is functional in the sense that remedy is suggested. It is distinguishable from the first only in the sense that it utilizes formal, agreed upon and presumably validated categories, often categories that are supported by considerable research and laboratory data. Such a diagnosis might be "myocardial infarction" or "lymphoma". These diagnoses, in Sober's terms, appear to us to be more quantitative, backed heavily by scientific data of all kinds, as opposed to the former diagnoses which seem more qualitative and clearly intuitive.

There is a third kind of diagnosis which, in my view, is not a diagnosis at all for clinical purposes. It relies heavily on research data, but the data are not of sufficient quality to be useful clinically. Moreover, the diagnosis does not directly prescribe treatment (or lead one conclusively not to treat). Nevertheless, such diagnoses serve a function in research by enabling us to categorize people with apparently similar symptomatologies. Indeed, I would characterize them as *research diagnoses*, diagnoses that are, as it were, aborning for the clinical world. I distinguish them sharply from those that are useful for clinical purposes.

Among such diagnoses are those that are used in psychiatry. G. Engel [6] has recently noted that psychiatric diagnoses are generally less valid than diagnoses from other areas of medicine. There seems to be a threshold for scientific data regarding a presumed syndrome or disease entity, which when achieved consigns the diagnoses we do make to a research status. The present status of data regarding categorical psychiatric diagnoses suggests to me that such diagnoses have not yet achieved the threshold that would make them useful for clinical work. Let me suggest some reasons why.

Agreement among diagnosticians. In order for a diagnosis to be clinically useful, there should be a high level of agreement among competent diagnosticians regarding that diagnosis. Of course, there are the occasional instances when a single diagnostician constitutes a majority of one against the judgments of colleagues. And this is especially the laudable case when the diagnosis can be verified independently. But when the overall level of agreement on run-of-the-mill cases is low, we have prima facie evidence that diagnosticians are not communicating with each other. Moreover, the courses of treatment which are likely to arise from such divergent diagnoses are themselves likely to be diverse, useless, if not downright dangerous in the clinical sense.

While there are quite a number of reports which indicate that the reliability of psychiatric diagnoses is low, such reports are of variable quality and will likely raise in the minds of objective observers, questions regarding prior bias of the researcher. But I submit that even data emerging from the efforts of those who are committed to the notion and viability of psychiatric diagnosis in clinical practice, leave much to be desired. For example, the results of six separate studies of reliability of diagnosis have recently been summarized by authors who are favorably inclined towards the current diagnostic paradigm [15]. If questions of bias emerge in this analysis, the bias will favor current procedures, rather than oppose them. The kappa values for these six studies are summarized in Table I.

TABLE I

Kappa Coefficients of Agreement on Broad and Specific Diagnostic Categories from Six Studies (from Spitzer & Fleiss, 1974)

Category	I	II	III	IV	V		VI	Mean
					New York	London		
Mental deficiency				.72				.72
Organic brain syndrome	.82	.90					.59	.77
Acute brain syndrome				.44				.44
Chronic brain syndrome				.64				.64
Alcoholism					.74	.68		.71
Psychosis	.73	.62		.56	.42	.43	.54	.55
Schizophrenia	.77		.42	.68	.32	.60	.65	.57
Affective Disorder					.19	.44	.59	.41
Neurotic depression			.47		.20	.10		.26
Psychotic depression				.19	.24	.30		.24
Manic-depressive				.33				.33
Involutional depression			.38	.21				.30
Personality disorder or Neurosis	.63			.51	.24	.36		.44
Personality disorder			.33	.56	.19	.22	.29	.32
Sociopathic			.53					.53
Neurosis		.52		.42	.26	.30	.48	.40
Anxiety reaction			.45					.45
Psychophysiological reaction				.38				.38

Observe first that this table provides data on common diagnoses. Rare or obscure syndromes, about which there might naturally be disagreement, are excluded. Second, although the table provides data on some common diagnoses, many are excluded. It is much rarer, for example, to find a diagnosis of "schizophrenia" than it is for the diagnosis to be differentiated according to its subtype, e.g., paranoid schizophrenia, simple schizophrenia and the like. Evidence from other studies indicates that reliability of these subdiagnoses is quite low. The point is not academic. These differential diagnoses are alleged to relate to treatment and prognosis. If agreement cannot be achieved for the diagnosis, what shall we say about treatment and prognosis?

Finally, these studies were conducted under near-optimal conditions. "Most of the studies involved diagnosticians of similar background and training, a factor that one would expect to contribute to good reliability. In addition, in some of the studies, special efforts were made to have the participant diagnosticians come to some agreement regarding diagnostic principles prior to the beginning of the study" [16]. One study conducted by Spitzer et al and cited in Spitzer and Fleiss [15], involved active cooperation between the diagnosticians. "No attempt was made to prevent the admitting therapists from discussing his diagnostic formulation with the supervising psychiatrist. It is assumed that such discussions often but not invariably took place" [15].

Despite favorableness of circumstance, these coefficients of agreement are indisputably low. They range as low as .10 for psychotic depression – not an uncommon diagnosis. The highest *average* coefficient (and whether one ought average these coefficients is highly dubious) is .77 for organic brain syndrome. Most of the average coefficients range between .24 (psychotic depression) and .55 (schizophrenia), surely not a sterling performance for a series of studies conducted over the past two decades. "One can only assume therefore, that agreement between run-of-the-mill diagnosticians of different orientations and backgrounds, as they work in routine clinical settings, is even poorer" [16].

Optimal reliability. The studies summarized above are consistent with many other studies, dating as early as 1938 [2], that indicate that the reliability of psychiatric diagnoses is terribly low. This comes as no news to those who are familiar with the literature in this area. Indeed, in a classic paper on the same subject, Zigler and Phillips [18] made the same point more than a decade ago.

Given this state of affairs, it is nevertheless fair to ask whether there is any evidence that reliability could be improved. Under optimal conditions, under the most ideal ones, is it not possible that sufficient advances might be made in the reliability of psychiatric diagnoses to suggest greater *promise* for such diagnoses than is presently available?

Fortunately, there are such studies. In one that was perhaps most favorable for incrementing reliability, a group of researchers who had devoted many years to the problems of diagnosis created a set of Research Diagnostic

Criteria – not the usual criteria that are used in ordinary practice, but criteria that were definitionally sharpened and for which the broad range of hidden meanings had been blunted by this group's long association with each other. These criteria were applied, not to patients whose behaviors are often ambiguous, but to case records of patients who had once been hospitalized. Case records tend to inflate coefficients of agreement, since they do not "move" or change as real patients do, and because the symptoms have already been selected for consideration by the person who wrote the record. The results are presented in the first three columns of Table II.

TABLE II

Kappa Coefficients of Agreement for Major Diagnosis by Various Pairs of Raters on 120 Case Records. (Modified from Spitzer, et al, 1975)

	Diagnostic System Being Used			
	Research Diagnostic Criteria			DSM II
	A vs B[a]	A vs C[a]	B vs C[a]	D vs E[a]
Schizophrenia	.78	.84	.84	.48
Other Schizophrenia	.68	.82	.70	.49
Schizo-affective	.22	.54	.24	.37
Affective Illness	.70	.74	.70	.48
Manic Illness	.82	.92	.76	.37
Major Depressive Illness [b]	.67	.74	.66	.25
Minor Depressive Illness [c]	.56	.66	.56	.07
Other	.69	.70	.74	.61
Briquet's Syndrome	—	—	1.00	—
Antisocial Personality	—	—	.66	—
Alcoholism	1.00	.66	.66	—
Drug Dependence	1.00	1.00	1.00	—
Obsessive-Compulsive Neurosis	.64	.52	.65	.42
Anxiety Neurosis	.65	.59	.52	.06
Phobic Neurosis	.66	.39	.66	.32
Borderline or Atypical Psy.	.24	..06	.20	—
Undiagnosed Illness	.45	.30	.59	—

[a] "A" is an eminent diagnostician and diagnostic researcher. "B" and "C" are his research assistants "who had considerable on-the-job training in making psychiatric ratings and in the use of RDC, "D" is a different diagnostician and diagnostic researcher, also eminent. "E" is one of 16 experienced American psychiatrists.

[b] In this study considered equivalent to any psychotic depressive illness in DSM-II.

[c] In this study considered equivalent to neurotic depression in DSM-I.

A dash indicates that either the diagnostic category was not available to one or both raters, or that none or only one of the raters ever used the category.

These results have been described as indicating high reliability for all of the major categories, and reliability coefficients generally higher than have ever been reported. In my view, that description is insufficiently conservative. Except for "drug dependence" and, in one instance, "alcoholism," few of the coefficients exceeds the fairly liberal criterion of .85. Moreover, under such optimal conditions, it is important to note where reliabilities continue to be low. The range of reliabilities for schizoaffective disorder lies between .22 and .54; for major depressive illness, between .66 and .74; and for borderline or a typical psychosis, between .06 and .24. These are by no means uncommon diagnoses. I suspect that they are much more commonly used than "affective illness" or "schizophrenia" without a qualifier (and certainly more common than "other"!).

The final column of Table II compares the diagnoses offered by an eminent diagnostician with those accorded by (one of) 16 experienced psychiatrists. In this case, all are using DSM II on the same patient records. And despite the optimality of these diagnostic conditions, the results are simply dismal, ranging from .06 to .61 (for "other" – that ubiquitous category that continually seems to have high pulling power).

THE PROGNOSIS FOR TRADITIONAL PSYCHIATRIC CLASSIFICATION

Studies of reliability of psychiatric diagnosis, conducted under a variety of favorable circumstances, make clear that whatever its apparent attractiveness to some, psychiatric classification is not yet ready for the clinical use. And reliability studies conducted under optimal conditions discourage the belief that they will be useful for public consumption in the next decade.

While there are many studies that indicate low reliability, very few studies indicate *why* reliability is so low. Two studies, however, are quite illuminating. Zigler and Phillips [18] found enormous overlap between the signs and symptoms that were used to form one diagnostic impression and those used to form others. That suggests that regardless of the DSM or psychiatric textbooks, psychiatric diagnosticians are going "their own way". A study by Ward, et al [17] is even more illuminating in this regard. In that effort, two psychiatrists conferred and discussed the reasons for diagnostic disagreement on 40 patients. Inadequacy of the diagnostic nosology accounted for 62.5% of the reasons for disagreement, while another 32.4% was accounted for by inconsistency on the part of the diagnostician. The patient was relatively stable: His inconsistency accounted for only 5% of the total disagreement. With nearly two-thirds of the disagreement attributable to the nosology, there remains little to commend the *Diagnostic and Statistical Manual* for clinical use.

The fact that reliability is low in other medical areas offers little comfort to the psychiatric diagnostician, both because the logic of that solace is absurd, and because other evidence is commonly available for medical diagnoses. And precisely because other evidence is available to confirm or disconfirm, one can tolerate a lower level of reliability [4]. Thus, if less than ideal reliability accrues to the interpretation of X-rays [3] or electrocardiograms [5], neither does the full burden of diagnosis rest on those techniques [8, 9]. One can palpate, check fever, examine urine, perform blood tests, and more, in order to corroborate or disconfirm the tentative diagnosis made on the basis of X-ray or EKG. But psychiatry has a very narrow data base. Psychiatric disorder cannot be identified from blood, urine or feces, under the microscope, on palpation, or from biopsy. Nor can one confirm or disconfirm a psychiatric diagnosis by "looking inside". In clinical psychiatry, one has only the patient's behaviors – what he or she says and does. And comparatively, that is a narrow base indeed, one that argues strongly for caution and conservatism, for powerful evidence of reliability and validity.

Umbrella effects. While coefficients of agreement among diagnosticians with regard to most functional psychiatric categories are low, the evidence mounts that there is such an entity as schizophrenia. As I have indicated, I would accord that entity the status of a research diagnosis, not merely because the diagnosis is presently unreliable (regardless of how well supported in the scientific literature) but also because the nature of the supportive data contributes little to clinical diagnosis.

The bulk of the evidence for the entity called schizophrenia derives from the genetic studies conducted by Rosenthal, Kety, Mednick and their colleagues [13, 14]. On the basis of twin studies, and utilizing the techniques of population genetics, these researchers conclude that there is a genetic component to schizophrenia. This conclusion has now been widely accepted.

Data from population genetics are rarely useful for clinical diagnosis, since clinical diagnosis deals with the individual and population genetics by definition does not. But population genetics can intrude negatively on a diagnosis, particularly when the category is fuzzy. An interesting example of this umbrella phenomenon has recently been published by Reade and Wertheimer [11]. A fictitious case history containing behaviorial clues that could be interpreted as moderately consistent with a diagnosis of schizophrenia, including such matters as introversion and untidiness, but that would not be conclusive, was prepared in two forms. The forms were identical except that the control form read, "Mr. S has an identical twin, who 7 years ago joined the army and has visited him only once for a 2-week long stay". The experimental form read, "Mr. S has an identical twin, who 4 years ago was placed in a mental hospital with a diagnosis of simple schizophrenia." Clinical faculty and interns were asked to judge the probability that Mr. S was schizophrenic on a scale from 0 to 100. The mean probability was 39 for the control, and 66 for the experimental group. Medians were 40 and 68 respectively. The mean clinical experience was the same for both groups: 5.5 years.

As the authors point out, diagnosis should be made on the behavior of the patient, and on nothing else. A genetic predisposition to schizophrenia is not schizophrenia until it is expressed phenotypically. The magnitude of the bias suggests that this umbrella effect is likely to be operative even under conditions of minimal ambiguity, simply because the genetic data are so widely known and regarded.

Umbrella effects, as might be imagined, operate not only in arriving at a diagnosis, but color perception of the patient after the diagnosis has been made. This is especially the case when the syndrome is thought to be massive, i.e., one that has many ramifications. A stroke, for example, is a physiological event which occurs with massive complications. It is therefore commonplace for a particular disturbance to be misattributed to stroke simply because strokes do have such large systemic effects. Similarly, in the psychiatric domain, there is the natural inclination to attribute all manner of bizarre behavior to, say, the patient's schizophrenia because that diagnosis is so all-encompassing. Many a side effect from psychotropic medication has gone unrecognized as a result, and has rather been attributed to the schizophrenia.

From research to clinical diagnoses. What is it, then, that elevates a diagnosis from research to clinical status? What is it that makes the notion of clinical judgment useful and respectable with regard to diagnosis and treatment? I doubt that we will find a single factor, but rather a complex family of factors that makes for respectability in the best sense of that word. One of them is agreement among diagnosticians. Is the evidence, and are the categories, such that diagnosticians can be rationally compelled to one alternative over others? A second, which is predicated on reliability, is validity. Third, usefulness. We have not considered base rate problems here, but if everyone is diagnosed X (read: schizophrenia, for example) then the differential utility of that diagnosis must be zero. Fourth, the kinds of data that can be marshalled to support the case. Some data, as we have seen, are interesting but in no way disposative. They are more likely to create negative umbrella effects than substantively contribute towards accurate diagnosis. Insofar as one can draw data from a wide base, from a spectrum of technologies and evidence, and to integrate those data, to that extent are we more likely to justify a clinical, rather than a research category.

Stanford University
Stanford, California

BIBLIOGRAPHY

1. American Psychiatric Association: 1968, '*Diagnostic and Statistical Manual of Mental Disorders* (DSM-II)', American Psychiatric Association, Washington, D.C.
2. Boisen, A. T.: 1938, 'Types of Dementia Praecox – A Study in Psychiatric Classification', *Psychiatry* **2**, 233–236.
3. Cochrane, A. L., and Garland, L. H.: 1952, 'Observer Error in Interpretation of Chest Films', *Lancet* **2**, 505–509.
4. Cronbach, L. J., Gleser, G. C., Harinder, N., and Nageswari, R.: 1972, '*The Dependability of Behavioral Measurements: Theory of Generalizability for Scores and Profiles*', New York: Wiley.
5. Davies, L. G.: 1958, 'Observer Variation in Reports on Electrocardiograms', *British Heart Journal* **20**, 153–161.
6. Engel, G. T.: 1977, 'The Need for a New Medical Model: A Challenge for Biomedicine', *Science* **196**, 129–136.
7. Garland, L. H.: 1960, 'The Problem of Observer Error', *Bulletin of the New York Academy of Medicine* **36**, 570–584.
8. Gilson, J. C. and Oldham, P. D.: 1952, 'Lung Function Tests in the Diagnosis of Pulmonary Emphysema: The Use of Discriminant Analysis', *Proceedings of the Royal Society of Medicine* **45**, 584–586.
9. Koran, L. M.: 1975, 'The Reliability of Clinical Methods, Data and Judgments', *New England Journal of Medicine* **293**, 642–646 and **293**, 695–701.
10. Meehl, P. E.: 1954, '*Clinical Versus Statistical Prediction*', University of Minnesota Press, Minneapolis, Minnesota.
11. Reade, W. K., and Wertheimer, M.: 1976, 'A Bias in the Diagnosis of Schizophrenia', *Journal of Consulting and Clinical Psychology* **5**, 878.
12. Rosenhan, D. L.: 1973, 'On Being Sane in Insane Places', *Science* **180**, 250–258.
13. Rosenthal, D.: 1971, '*Genetics of Psychopathology*', McGraw-Hill, New York.
14. Rosenthal, D., and Kety, S. S. (eds.): 1968, *The Transmission of Schizophrenia*, Pergamon Press, London.
15. Spitzer, R. L. and Fleiss, J. L.: 1974, 'A Reanalysis of the Reliability of Psychiatric Diagnosis', *British Journal of Psychiatry* **125**, 341–347.
16. Spitzer, R. L. and Wilson, P. T.: 'Nosology and the Official Psychiatric Nomenclature'. In A. Freedman and H. Kaplan (eds.). *Comprehensive Textbook of Psychiatry*, Williams and Wilkins, in press, New York.

17. Ward, C. H., Beck, A. T., Mendelson, M., Mock, J. E., and Erbaugh, J. K.: 1962, 'The Psychiatric Nomenclature: Reasons for Diagnostic Disagreement', *Archives of General Psychiatry* **7**, 198–205.
18. Zigler, E., and Phillips, L.: 1961, 'Psychiatric Diagnosis: A Critique', *Journal of Abnormal and Social Psychology* **63**, 607–618.

SECTION II

THE LOGIC OF HEALTH CARE

EDMOND A. MURPHY

CLASSIFICATION AND ITS ALTERNATIVES

In a former communication [32], I have discussed three issues which bear on my topic[1] – classification, bimodality, and abnormality and disease. It is an elementary rather than an elemental text. In so sophisticated a group I feel that I can afford to dig a little deeper. At the same time, I should warn you that my point of departure will be strictly scientific: I shall be content to leave philosophy to the philosophers. My invasion into the fringes of epistemology is prompted because I have found repeatedly that the received questions which I have attempted to explore scientifically prove to be obscure or even meaningless; and in order to get back on the proper road of inquiry it is necessary perhaps to restate the fundamental problems.

I start with three simple but basic ideas.

First, systematization, the major endeavor of science, presupposes that the totality of all things can be described more simply and more economically than the data themselves. If it takes n formulae to describe the paths of n planets, then in the economy of science nothing has been gained. If there is one formula, Newton's inverse square law, which describes all, then we have indeed made a scientific advance. Of course that does not make of Newton's law the ultimate truth; but to a first order of approximation it has great utility. This idea of economization is so basic that I propose to call it the fundamental desideratum of science. But while all might agree with it, there is much debate as to how it is best pursued. Laplace surmised that if we knew the position and velocity of every particle in the universe we could reconstruct the past and predict the future exactly. However for a finite deterministic universe this surmise could never be tested empirically (see Appendix). For an infinite universe the proof is more complicated. The statement applies *a fortiori* if, as modern physics suggests, there is a random element in nature. Thus, belief in the feasibility of such a reduction must be *a priori*. Even so, we may hope for, but may not demand, simplicity.

My second point is that if truth exists, (and by "truth" I mean that which exhibits universal invariance[2] and which transcends the imperfections of individual observers), then it is difficult and perhaps impossible to attain. We talk in biological sciences of "facts" which, if sophisticated enough, we distinguish from inferences and opinions. But deeper analysis suggests that they

H. T. Engelhardt, Jr., S. F. Spicker and B. Towers (eds.), Clinical Judgment: A Critical Appraisal, 59–85.

might better be called artifacts. What we conceive, rather naively, to be the raw data of experience are inextricably mixed up with our preconceptions. A fact is, in the strict sense, an abstraction: that is, it is an incomplete selection of those things with which the senses come in contact; and it is not particularly surprising that others making different selections may attain different facts. It is commonly said that seeing is believing, as if experience were always anterior to fact. But just as often the converse is true: believing is seeing. Sullivan writes: "Faced with the amazing number and variety of impressions we call Nature it is only by great effort that man has learnt to discriminate those that may be successfully used for constructing science" [39]. He adds that radically different viewpoints may have utility. Of course the intrusion of the subjective applies to the perceptions of the mind as well as the data of the senses. And a major problem, even in scientific inquiry, is the false necessity of thought. In some degree, these dispensable convictions are cultural. The idea is an old one. "Custom, and especially custom in a child, comes to have the force of nature. As a result, what the mind is steeped in from childhood it clings to very firmly, as something known naturally and self-evidently" [2]. Bronowsky [10] points out that the most revolutionary axiom of relativity physics is that the observer is part of the system and hence that his perceptions of it are incurably colored by his own framework of reference. This rather obvious point suggests that the most appropriate method of science is iterative. We have rather suffered from the Greek approach to inquiry, so admirably demonstrated in Euclid and expressed in Lewis Carroll's phrase: "Begin at the beginning and go till you come to the end; then stop" ([11], p. 143). Nothing encourages the view that we have any right at the start to lay down any unassailable conceptual framework which later we may not have to modify or even discard. Exposition should be orderly; but the idea that *inquiry* is a coherent and systematic progression is pernicious. There is nothing in this of romantic pessimism. Some three hundred years of assault on the quintic equation culminated in Abel's celebrated proof that it cannot be solved systematically. But this has not led to despair. If the equation cannot be solved systematically then it can be solved (somewhat inelegantly) by iteration: that is, by continually modifying conjectured solutions in accordance with their demonstrated inadequacy. It may some day be proved that the whole scientific endeavor is by its very nature inaccessible. It may be impossible to give a completely coherent description of the universe. Nevertheless science may still be able to solve each particular problem to any desired degree of accuracy. Such a conclusion would be a disappointment to the academic mind; but that is no reason why it should lead to black despair.

My third point which is of no interest to the scientist as such, but of cardinal importance to the physician, is a spiritual one: a deeply-rooted belief in, and reverence for, the individuality of the person. Physicians with sensitivity do not talk about "cases of pneumonia" but of "persons with pneumonia." This admirable sentiment is in some danger of degenerating into a mere attitude: a belief that even to attempt a rigorous scientific formulation of the characteristics of the patient is a kind of blasphemy. It results from a profound confusion between the economy of characterization possible with even quite simple descriptors and the infinite repertoire of individuations. Nevertheless, we must respect this desideratum of uniqueness.

The relative merits of classification and individuation constitute a very old problem. Unfortunately, so much of what we call medical science has attempted to canonize as fundamental truths what are only makeshift devices invented in the face of harsh practical necessities. Of course, medicine is also a practical art and there is an enormous resistance (which is easy to understand if not to condone) to calling these so-called truths into question. There is a wide belief that there are things called diseases which it is the business of the physician to detect and cure. I would not flatly say that this whole way of looking at medicine is a delusion. But I have expended much effort in vain in attempting to provide a coherent basis for it. So far I have no definition of disease which is proof against absurdity in individual cases. Now rather obviously the notions of "health" and "disease" conform to the minimum requirements of a coherent system of classification. Thus we would have a practical case against which we could gauge the utility of classification. But I will not follow this now familiar path. Instead I wish to explore the problem of classification in more general and fundamental terms.

THE USES OF CATEGORIZATION[3]

I shall use the term "categorization" to mean the grouping of elements into classes or categories where the process is a deliberate imposing of a system not generated from the empirical data but from some principle of unity conceived in the mind of the scientist. It is to be distinguished from clustering where the criterion is one of proximity in a metric space (see below).

There are three major issues I would first like to discuss: scientific economy, the message-noise dichotomy, and the conclusions of finite experience.

Scientific Economy

One of the motives for classifying (for example, classifying patients according to the diseases from which they suffer) is to reduce complexity. Most physicians I fancy would be happier with the statement that Mrs. Jones has pneumonia than with a recital of her physical manifestations and laboratory findings. The physician has seen "pneumonia" before and will feel the patient must have a certain similarity to the others; on the other hand he may never have met Mrs. Jones before. He may be uneasy juggling dozens of facts in his mind rather than seeing Mrs. Jones' particular needs in terms of a concrete problem which from experience he can handle. Of course such a view of disease always breeds something of a reaction begotten of the desideratum of uniqueness: a stressing of Mrs. Jones' particular needs and idiosyncrasies. The physician will perhaps note her unusually high white cell count, or her severe anemia. But such features are probably commonly regarded as modifying a substantial categorical disease.

The Message-Noise Dichotomy

Consider three pieces of music: Mozart's "Hunt" string quartet; a romance for bagpipe and orchestra; and John Gage's piece *4′33″*. I hear them on my short wave radio set which emits hums and crackles, and must try to separate the significant music which the composer intended, from the idiosyncrasies of my radio which he did not. The objective of the enterprise is clear enough; but the problems are vastly different. Mozart's idiom is restrained and predictable. He did not use non-musical sounds nor *ostinati*. If I hear a hum, it must be extraneous. With a romance for bagpipes it is not so clear. If I hear a hum can I be sure that it is in my radio and not part of the music itself? The third example is at once much simpler but also more bizzare. For the work by Gage comprises a prescribed period of uninterrupted silence. *Any* sound that we hear must be noise. As a reference we might add a fourth category of sound: whatever is produced on the ideal radio when any one of these pieces is played. The first three instances carry over by analogy to scientific data where the underlying relevant information, the message, is well understood and predictable (Mozart); where the existence of the message is clear but the criteria for educing it uncertain (the bagpipe); and where there is no message at all (Gage). We compare them with the fourth class where there are no faults in our perceptions, only in our conceptions.

A biologist might argue that there exist substantial groups in nature, known to us more or less imperfectly by their manifestations and our observations. Roughly speaking this view corresponds to the Aristotelian idea of substance and accidents, but with stochastic, that is random, overtones which, particularly in the world of biology, loom large. We postulate the class, and suppose that because of inattention, or random error, or the inclusion of other irrelevancies, we discern it fallibly and with difficulty. The object of classification is, then, to recognize the existence of these substantial classes explicitly, and to devise the most efficient and reliable tactics for distinguishing among them by discriminant analysis.

By way of an illustration consider the categorization "sex." It is widely held that there are two categories of people in the human species, male and female. A great many scientific criteria have been devised: anatomical, psychological, physiological, biochemical, cytological, among others. All of them are at one time or another ambiguous; and they are certainly not always mutually consistent. But we continue to believe that there exist males and females and this truth we could call (by analogy) "message." In any individual case the data presented to us may be erroneous and misleading. For example consider the cytogenetic evidence. In the epithelial cells of the mouth the female should have a particular structure, the Barr body, in every cell whereas the male would lack them. But the counts obtained by this method are not always either 0% or 100%. For the most part they assume intermediate values which for purposes of decision we separate into two

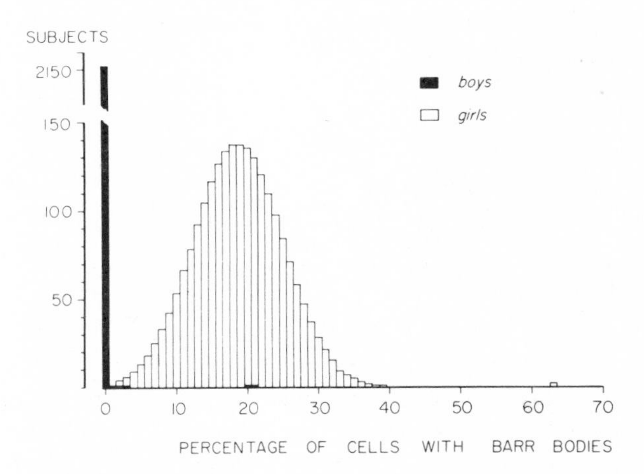

Fig. 1. Histogram of the percentage of cells with Barr bodies in an unselected laboratory series. The black cluster of values to the left represents males, the white cluster to the right females. The separation is clear; but the two populations are not uniquely located at the two hypothetical points 0% and 100%. (Reconstructed from the data of [42]).

clusters (Fig. 1). The observed departures from the values 0% and 100% represent various extraneous factors including observer error which tells us nothing about the underlying truth, merely something about the imperfectness with which we apprehend it, and hence "noise." Examining epithelial cells is a worthwhile endeavor which we must not belittle simply because occasionally we make mistakes in our discrimination.

Is it fundamentally unreasonable that conclusions may attain a resolving power transcending that in the data on which they are based? Let me give you an example which I think is to the point. According to Dalton's atomic theory the atomic weight of a metal equals the product of two characteristics: its equivalent weight which is easily found and its valency which is not. It was discovered empirically by Dulong and Petit [17], that the product of the atomic weight of a metal and its specific heat is approximately equal to 6.4. The specific heat is determined easily enough and when divided into 6.4 should give an approximate atomic weight. Now the valency must be a whole number. So we take as the atomic weight that exact multiple of the equivalent weight closest to the Dulong-Petit estimate. By this means we can start with an approximate atomic weight and end with an exact one (Table 1). In both cases, the use of the Dulong-Petit law, and the criterion of maleness, the only quantity that we are interested in, is an exact answer. The departures from this exact answer have no significance and the information on them could be discarded as pure noise.

TABLE I

The Use of the Dulong-Petit Law (DPL)

Metal	Specific Heat at 25°C	Atomic Weight estimated from DPL	Equivalent Weight	Estimated Valency	Inferred Valency	True atomic weight
Copper	0.092	69.57	31.773	2.19	2	63.546
Lead	0.038	168.42	103.6	1.63	2	207.2
Zinc	0.0928	68.97	32.69	2.11	2	65.38

Now indiscriminate use of these arguments gives rise to uneasiness over several points. In the chemical example, the major point is essentially axiomatic. If Dalton's theory were totally false the bottom would fall out of the entire argument: there would be no particular reason for example why the valency should be a whole number. On the other hand there is a fair mass of collateral evidence, comprising what are commonly viewed as deductive tests [33] though many are perhaps inductive, that Dalton's theory is true.

The atomic theory of matter stems from a critical issue of choice between an atomic and a continuous view of the physical universe. Democritus had explored this problem out of an intellectual necessity before there were any precise chemical data to prompt it at all. However it is not at all clear that sex is a comparable issue. There does not appear to be *a priori* any reason for demanding that there should be only two sexes in the human species; and the belief is based on purely empirical observation. It is by no means impossible to imagine things otherwise. And it seems clear that there is a difference between a construct based on empiricism and the evidence of our senses, and a construct which, perhaps, we have to authenticate empirically, but which is at least prompted in the first place by logical analysis. If matter consists of particles, then the elemental interactions among them must involve whole numbers. We cannot abandon that belief without discarding our whole concept of a particle. Then we can attach no meaning to a valency of 2.3 and we are then more or less forced to conclude that the quantity ".3" can have no meaning for the fundamental interactions which we are studying. (It might, of course, represent an average of disparate integral quantities.) But suppose that in a patient ten percent of the cells contain Barr bodies. Such a figure is sound evidence that the patient is female. Are we justified in taking the further step of saying that the figure 90% (the deviation from the ideal theoretical figure of 100%) has no meaning whatsoever?

Often the issues are even less clear cut. If the clinician divides his subjects into those with, and those without, coronary disease on the basis of some measurable quantity, does that mean that he wants only to categorize? Is all the other information completely irrelevant? Are there no *degrees* of coronary disease? Or would he say rather that there are many gradations of change in the coronary arteries only some of which cause overt harm? Certainly the latter attitude explicitly proposed by White twenty-six years ago ([43], pp. 38–40) could be much more readily defended than the former. It might be argued that the idea of coronary disease is not ineluctable but that it is a factitious phenomenalization which we have imposed on what we have, or think we have, seen.

There are yet more flagrant areas of doubt and ambiguity where we adduce no theory that the manifestations are a true but hazy outline of some inner reality: where instead we are attempting to define the state on the basis of measurement alone. This means that the diagnosis is (tautologously) infallible. The price for this facile success is uncertainty as to whether the categorizations have any meaning which transcends the original data; and that certain strange and even, occasionally, absurd conclusions may result. The most obvious problem arises in connection with scientific discussions about intelligence which in fact is not being measured but *defined* by intelligence tests.

Constraints of Finite Experience

The theorist – the mathematician for instance – can commonly call on many examples of his problem e.g., prime numbers. The empirical sciences do not have infinite amounts of data; and even large samples may be prohibitively expensive, or (especially in biology) maintaining uniform standards of observations may impose logistical problems. True relationships may be continuous but any possible set of individual observations is necessarily discrete. It is analogous to ordinary vision which is mediated through a finite number of retinal cells and which therefore is dotty like the paintings of Pissarro, and which the visual cortex represents as a spatially continuous perception.

For practical purposes, and as a means of making general conclusions even about continuous characteristics, the scientist may see fit to group his finite number of data arbitrarily. For instance in the Down syndrome (or trisomy of chromosome 21) there is interest in the impact of maternal age at conception. If the mothers are put into five-year age groups, the incidence of this chromosomal abnormality proves higher in the progeny of the older age groups than in the younger. The investigator implies no belief that the *boundaries* between the groups into which the mothers have been divided have in themselves any intrinsic scientific meaning. Maternal age might be grouped in various ways (Fig. 2) and while details would differ, there would be broad invariance of the *trend*.

Parenthetically, I may point out that those who use methods involving arbitrary grouping have little control over the misuse that may be made of the results by those with less clear ideas. The latter may feel that the dividing lines (which were chosen arbitrarily in the first place) actually have some real meaning. They will then carelessly talk about such classes as "hypertension" which are then reified.

The difficulties of the basic scientist are aggravated when he wishes to look at the joint effects of several factors. Even a comparatively parsimonious grouping may produce a very large number of classes. With samples of even the maximum size practicable, there may be embarrassingly few subjects in each group. Thus the tendency is always to group coarsely and to tolerate much heterogeneity within classes. The remedy, larger samples, may reduce

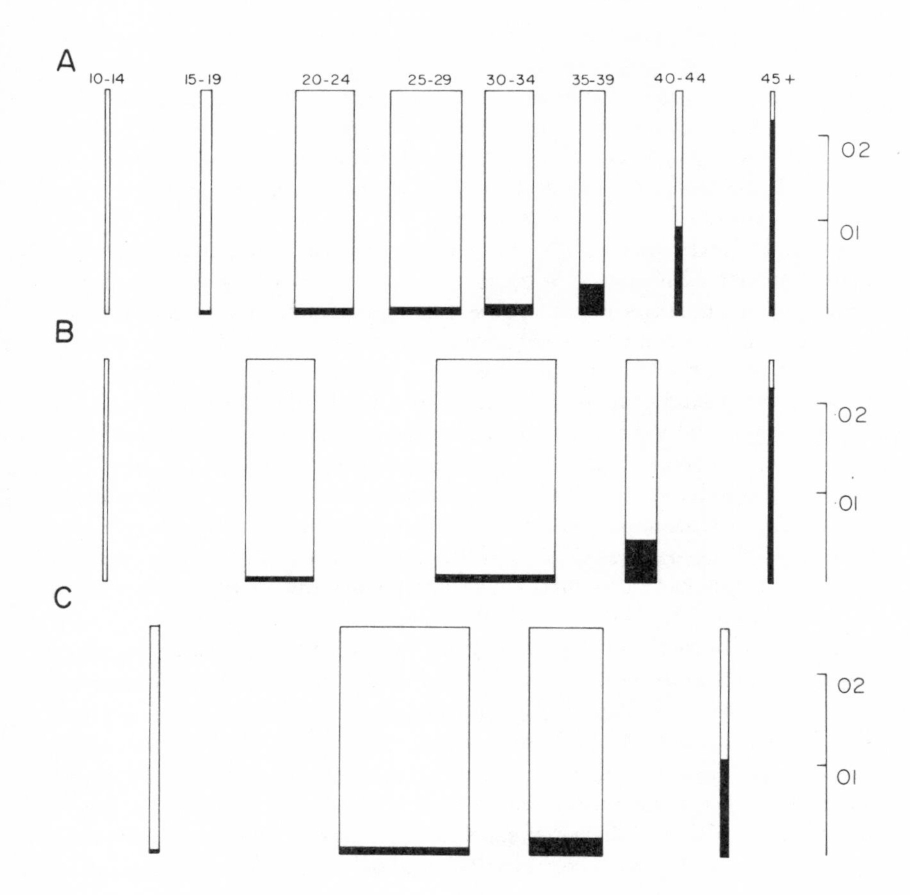

Fig. 2. Frequency of Down syndrome in relationship to maternal age. A. Grouped by five-year age groups. B. Regrouped into ten years age groups. C. An alternative regrouping. The *width* of each cell represents the number of mothers involved (which gives some indication of the accuracy of the estimate). The *height* of the black part represents the proportion of children affected. Pooled data are represented midway between the groups pooled. The three groupings give different results; but in all three there is a clear increase in frequency with maternal age. (Data of [15]).

heterogeneity by allowing finer grouping; but it will introduce other heterogeneity due to observers, to non-simultaneity in the observations, or some such.

THE CRITERION OF CLUSTERING

An alternative approach to imposing a grouping on the data is to let the data sort themselves out into groups or clusters. Mendel, for instance, found that peas comprise two groups, tall and short [29]. They breed true within groups in accordance with the laws of genetics. But this does not imply that short pea plants are all of exactly the same height.

We could tackle the somewhat more ambitious task of taking two measurements, plotting each individual point on a plane and looking for two-dimensional clusters, much as the ancient astronomers grouped stars into constellations. So long as astronomers perceived the night sky as revolving around the earth, the members of a constellation would appear to travel together like a flock of sheep, and perhaps support unconsciously the idea that somehow they belong to each other. But now we no longer think of them as traveling in step; since, moreover, it is only with comparatively sophisticated apparatus that one can judge accurately the distance of stars, the stars in a constellation may in fact be separated by vast distances and have only in common that they are located in a particular quarter of the heavens as viewed from the earth. A map of the stars in three-dimensional space might show that the clustering of stars was illusory (Fig. 3).

In yet more general terms the taxonomist of human disease can make more and more measurements, plot them, as it were, in conceptual spaces of higher and higher dimensionality and find out whether patients, suffering from unidentified disease, fall into clusters.

There has been a tendency to suppose that such systematic methods of cluster analysis are the answer to our problems. However, I have the intuitive impression that the fundamental cause of the failure is that we are trying to make statistical methods solve problems beyond the scope of statistics and its axiomatic underpinning. It is as if we were to attempt to base a tasteful color-scheme on Euclid's geometry and its axioms. The specific problems are as follows.

First there must be some prior warrants. A sufficiently free analysis of fifty data points would conclude that there are fifty clusters, each containing one point only. This result certainly guarantees homogeneity within groups. But it ignores the fundamental scientific desideratum. Converting fifty individual data into fifty classes is not a reduction. Now we may constrain the solution so that there are (say) five clusters. But if *a priori* we do not know how many clusters we are looking for, why, clearly, we have no basis for choosing five rather than any other number. We might set up a rate of exchange between

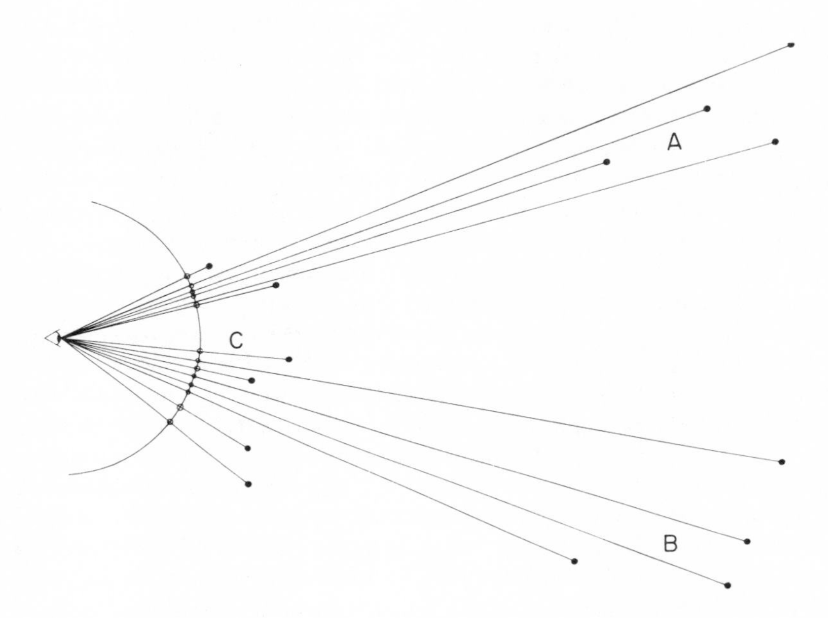

Fig. 3. The impact of position on the way in which the observer discerns clusters. We suppose that observer cannot distinguish distances, only angles. The diagram shows three clusters of stars (black dots), one comparatively near the observer (C), two more distant (A and B), all well separated. But the angular distribution as the observer sees it (represented by the white dots on the circle) shows two clusters, each a mixture of near and distant elements.

the amount of heterogeneity within clusters and the number of clusters (both of which quantities we wish to be small). But what rational basis have we for such a rate of exchange? I find it very difficult to believe that it can be derived from the data. Indeed I think there would be general agreement that values cannot be demonstrably educed from brute fact.

Then, second, the results are scale-dependent. If all the measurements were of the same type – for instance, all were special distances – there would be no special problem in the choice of scale: all would be measured in the same units. Even so, it is not clear that the lengths of noses and of freeways should both be measured in (say) meters. But this is, after all, only an incommensurability about meaning. Where they are also physically incommensurable – where one measurement is a body temperature and one a white cell count – who is to say what their relative importances are? Of course if the issue is discriminant analysis, if there are two known groups, those with, and those without pneumonia, the process of optimizing the discrimination will itself take care of the scaling. But the problem is that we may have no scientific basis for assuming that the groups exist or, if they do, how many they are. Still less is there an infallible reference criterion from which the weightings may be estimated and against which the methods can be validated. Of astronomic measurements two, the coordinates of the star, are angular; the third a linear distance. As the primitive astronomer was ill equipped to measure distance it was played down as a criterion of clustering, whereas angular differences are easily measured and therefore received much emphasis.

The third, and perhaps the most difficult, problem lies in the choice of a metric. In particular, we have two problems, linearity and monotonicity. In measuring height, we think in linear terms. A man six feet tall is twice as tall as a child three feet tall. But many measurements are not made on such a scale. The level of sound and the intensity of an earthquake are measured on logarithmic scales. Momentum is measured in terms of velocity, kinetic energy in velocity squared. Lenses may be measured in focal length, which is linear, or diopters which are on the reciprocal of a linear scale. We may size red blood cells by diameter, which is linear, or mean corpuscular volume, which is cubic. There is a long-standing controversy over why empirical numbers in any field of scientific endeavor are more likely to begin with the number one than the number nine [34, 35]. For my part, I think that this phenomenon reflects how we think of physical phenomena: we lump the Earth and Venus together as planets (although they differ by many trillions of tons); on the other hand we call the hydrogen atom tiny, and the uranium atom enormous, although the absolute differences are minute. Yet it is doubtful that what we group together as atoms, insects, trees, planets, stars, galaxies (because *we* think in *orders* of magnitude) tells us anything about nature.

In medicine, we have to deal with even less well resolved questions. A colleague once remarked that our administration labors under the delusion that two people each with I.Q.'s of 50 are equivalent to one person with an I.Q. of 100. The equivalence would be absurd; but it is an illuminating question *why* it is absurd.

This discussion may appear much ado about nothing. But even for a single measurement there may be factitious evidence of grouping if a non-linear relationship is used [5, 31]. Whether we decide there are two populations of red cells in a patient may depend on whether we measure the diameters or calculate from them the volumes of the individual cells. But it is absurd that two such opposing conclusions should depend on purely arbitrary manipulations of the same data. In science we look for greater invariance than that.

But in the examples I have given so far, we are assured of monotonicity: at least the ranking of magnitude would be invariant. Ten is greater than one; and the roots, the powers, the logarithm, the antilogarithm of ten would all be greater than the corresponding function of one. But neither in mathematics nor in biology do we necessarily have that assurance. The sine of an angle is a periodic relationship. Patients with heart failure may vomit either because they have not had enough, or because they have had too much, digitalis. Patients may be in peril of their lives because they are too fat or too thin. Thus vomiting is not a monotonic function of dosage of digitalis, nor prognosis of body weight. The impact of heparin on platelet survival in the dog [38] exhibits at least four alternating phases of long or short survival as the dose is increased. It clearly will not do to group patients by some measurement without assurance that low values always mean more or less the same thing and conversely. The instances quoted are obvious and comparatively well known. What assurance then, have we, that precisely the same process is not at work in processes about which we know much less than in the examples quoted? But a disembodied analysis of data based on the bland principle that a measure is a measure is a measure, cannot meet such an objection.

Let me add to these three, a fourth problem which will pave the way for the next section. Are some measurements pure noise in some context? Implicitly, this supposition is often made, if only by default. The scientist rarely attempts to accumulate all the information that is in principle extractable from any event, however precious. Amassing information is always selective; and by ignoring the rest, we usually imply that they are irrelevant. But this judgment reflects most immediately not what the world is like, but the way in which we see it. The great Ramon y Cajal [36] described the Barr body independently of Barr [6] ; but it never struck him that it occurred in female cells only.

DIMENSIONALITY AND CARDINALITY

There is yet a third approach which we may take, which was foreshadowed in the previous section. To expound it I must distinguish between the cardinality of the set and the dimensionality of the system. I have six symbols on the license plate of my car: two letters followed by four digits. There are 676 combinations of two letters and ten thousand four-digit numbers. Thus altogether there are 6,760,000 such license plates possible. The dimensionality of the system is six, the number of symbols on my license plate; the cardinality of the set of all possible such license plates is 6,760,000. A high degree of individuality may be observed among cars which can be attained by the use of only six symbols. If height were measured accurately only to one inch there might be some sixty groups of people from that one measurement alone. If we were now to introduce another measurement, such as the distance between the eyes, it would be possible to distinguish many people who had the same height, by the difference in interocular distances. The Bertillon system of anthropometry [17] which antedated the use of fingerprints, attempted to identify each person individually by a series of such measurements. It is a common cry among epidemiologists that extensive grouping whether arbitrarily imposed or arising naturally from the data, rapidly leads to an embarrassingly large number of groups and very few observations in each. The rational alternative then is to abandon the use of grouping altogether. How is this to be done?

REGRESSION ANALYSIS

To study the relationship between weight and blood pressure the categorizer would put the patients into (say) three groups by weight and three by blood pressure. If those with the lowest weight had the lowest blood pressure and conversely, he would infer that the two are related. But regression analysis proposes to deal with the issue by constructing a graph ("regression line") showing the relationship between weight and average blood pressure where no groupings are made at all and the characteristic constants of the regression line are to be estimated from the data. Thus a full use of the individual measurement is made and no information lost by gratuitous grouping. Most commonly a straight line is fitted; but lines of many particular mathematical forms can be used. Such a line gives a best estimate e.g. for predicting average length of life for any particular weight (Fig. 4a).

Regression analysis and its elaborations have become popular particularly among epidemiologists interested in such topics as cardiovascular disease. Its advantages are obvious and under certain conditions the resulting analyses are much more precise than those based on grouping. One is left, then, with the surmise that the old systems of classification have little use other than simplicity, and the sooner they are given a decent funeral and replaced by regression methods the better.

But philosophers (who take the long view of knowledge), will feel that there must be a catch somewhere. In the ultimate economy of things, we get nothing for nothing. Like the illustration of the Dulong-Petit law there must be some hidden source of logical support which makes two blossoms grow where previously only one could. In fact, I can identify three major difficulties.

In the first place there is a logical difference between the notion of a continuous relationship such as that between the probability of heart disease and the blood cholesterol level, and the idea that there really are qualitatively different classes of things. The philosopher who believes in the existence of coronary disease as a discrete entity with the same enthusiasm that he believes in sex, will have some difficulty with the notion of gradual transition from one state to another. Here etymology betrays us. I talk about a gradual change; the statistician talks about the gradient (i.e. the slope) of the line. Both notions technically imply a smooth continuous change. But both words are derived from the Latin word *gradus*, a step which is a discrete instantaneous change. The curve usually fitted relating incidence of coronary disease to blood cholesterol does not in the least look like a flight of steps in profile, but rather like a ski slope. And if the progression of coronary disease is not a slow, imperceptible, continuous process of decay but a series of minor catastrophes in which the blood vessels are instantaneously blocked off one by one, it is doubtful that this relationship is appropriately represented by a continuous smooth curve. A detailed discussion of this problem would get us into very deep water including an epistemological analysis of what we mean by probability.

The second objection to the use of regression analysis is closely related. There has been a tendency to define hypertension as a systolic blood pressure of greater than some particular value, let us say 145 mm Hg. And we say the prognosis in hypertensive patients is worse than in those who have normal blood pressure. For what it is worth there is no objection to this statement. But it may be rather absurdly construed to mean that between somebody whose blood pressure is 144, and therefore by definition normotensive, and somebody whose blood pressure is 146 and therefore by definition hypertensive, there is an abrupt and catastrophic change in prognosis. There is no empirical evidence to support this view; and, if the object is prognosis, the arbitrary drawing of such a sharp dividing line has little merit. But there is a new branch of mathematics (which has not yet invaded the traditional statistical field) known as catastrophe theory [44]. It embodies the idea that over very small changes in one characteristic there may indeed be a large and abrupt discontinuity in the behavior of the other. The champion of this theory of mathematics is Thom [41], a contemporary French mathematician, who has applied his model extensively to biological processes.

Regression analysis is characteristically performed in systems of noisy data so that we cannot reasonably expect the data to conform precisely to any curve fitted through the data points, however apt. Now some of the discrepancies may be due to noise; but equally they may be due to incongruity of the regression line which is being fitted (Fig. 4b). I know very little, if any, statistical research on discontinuities of this kind. There have been some

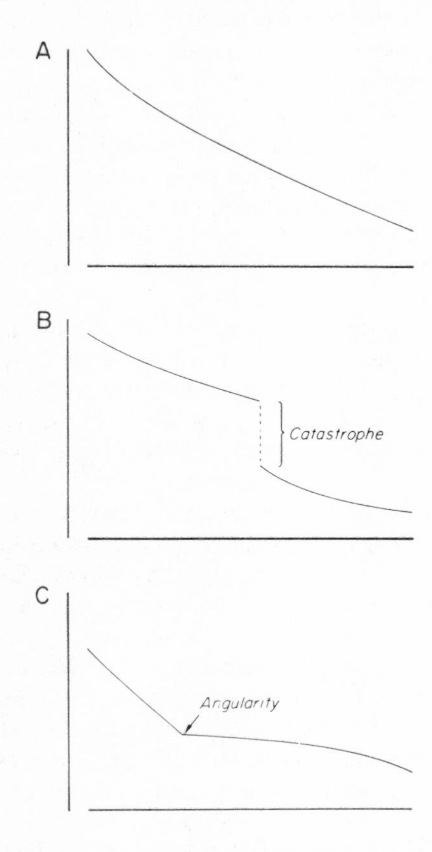

Fig. 4. Three types of regression curves. A. A continuous function with a continuous first derivative. B. A function with a catastrophe which is not continuous. C. An angularity or spline, which is continuous but does not have a continuous first derivative.

studies from time to time on "splines" or angularities, that is, points at which the first derivative of the function is not continuous (Fig. 4c). But even in such models, statisticians have been at pains to insist that there is at least continuity of the function itself. Thus their whole view of the world is calculated to obscure the catastrophes which certainly exist.

For instance, consider the relationship of the length of subsequent survival to the level of uric acid in the blood. This characteristic is influenced by many factors and in turn has a variety of impacts. It increases in kidney failure, for instance. It is at least empirically related to the risk of coronary disease [30]. Thus we expect a curve of relationship with a negative slope. But uric acid has low solubility and at a certain critical level it forms crystals which perhaps produce gout and its consequences, including kidney damage, the formation of renal calculi and so forth. Thus we might expect an abrupt change, a discontinuity in the prognosis in the neighborhood of this critical level of crystallization. To fit a smooth prognostic curve through the data would thus be quite inappropriate; and an angularity would be no better. What we must fit is a discontinuous function with an abrupt leap at one point (Fig. 5). Our common implicit argument that the continuity of the regression line may be used to allow the data points to be mutually correcting – an argument I myself have used elsewhere [32] – would clearly break down here. It might be argued that I have picked on an extraordinary and pathological case; but this is by no means evident. We know little enough about biological processes and how to construct statistical models of them. The more we retreat from harsh reality into models with seductive mathematical properties, the greater the risks. Biological phenomena may be riddled with catastophes of this type, which we are at present too ill-equipped to perceive.

The third weakness in regression analysis is in some ways the generalization of the forgoing and perhaps the most insidious defect of all. Numerical analysis shows that one can find approximations to complicated functions of a totally different form [24]. For instance a polynomial may approximate a transcendental function. Some of these approximations are extremely good and while that no doubt delights the mathematician it should terrify the scientist. For it is not the *prime* responsibility of the scientist to predict the consequences of a series of axioms; rather it is the converse – by analysis to infer from data what the underlying truths may be. It must be obvious that if the correspondence is degenerate or nearly so and very different sets of axioms may produce results which in the presence of noise are difficult to distinguish, the scientist may be at a loss to decide which of the sets of very difficult axioms corresponds most closely to the truth. By analogy the junior

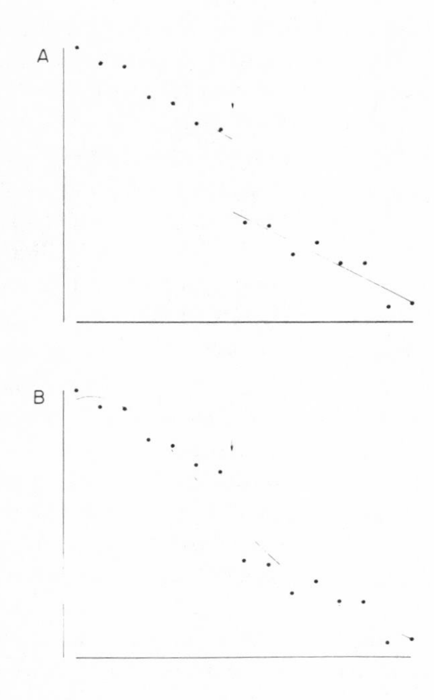

Fig. 5. A hypothetical example of a linear decay with a catastrophic drop at the arrow. Data points are produced by adding a random normal error to the expected value on the true decay curve. A. Two straight lines are fitted, two parameters (intercept and slope) being estimated for each, four parameters in all. B. The best fitting quartic curve fitted to the same point (five parameters being estimated). In the latter case the residual sum of squares is 3.82 times as large as in the former, although there is one less degree of freedom. The unprejudiced eye might suspect that something strange was happening in the neighborhood of the catastrophe; but this could easily be dismissed as an artifact by the statistician – a large positive error just before it, a large negative one just after it, both entirely due to chance. Certainly the fitted curve in the lower diagram looks plausible.

medical student learning to describe murmurs is doubtless comforted by the similarity between the murmurs of aortic and pulmonary incompetence since he has only one sound to memorize. But the clinician who discovers a murmur of this type in a patient may be perplexed by this very similarity because it is important for him to be able to distinguish which of the two types of lesions has produced this particular sound.

I have had a mounting preoccupation with the tendency of quantitative biologists in applying regression methods to have no concern whatsoever with justification other than that the fitted curves seem to fit the data adequately. Of course if the object of the regression analysis is merely a complicated form of interpolation, like the polynomial approximation for the mathematical function, then in some cases it may be useful. But if the issue is to make inferences about basic mechanisms, then what has been one of our best allies becomes one of our worst enemies.

For example, the temperature in degrees Fahrenheit is 1.8 times the temperature in degrees centigrade plus 32. There is an old and still useful formula that the mean systolic pressure in mm Hg for a patient is 100 plus the age in years. The formulae are similar in pattern but have totally different epistemological standings. The first is a completely rational adjustment for differences in two linear scales representing precisely the same phenomenon; and it is entirely appropriate even if there is error of measurement. It is known in statistical theory as a functional relationship [23]. But the second relationship is purely empirical. Blood pressure reflects many factors each of which in turn has many antecedents. We can make no useful detailed inference at all about underlying mechanisms from this formula, the only utility of which is descriptive.

I dare say that eventually we shall find that biological science is littered with spurious ideas which reify fictions of analysis. It is easy to see the examples which have been discredited – the philosopher's stone, phlogiston, gravitation, the ether. But ". . . men have insensibly fallen into turning an entirely negative term into a positive term. . . As if one were to talk of being on friendly terms with a gap in a narrative or the hole in an argument, of taking a walk with a non-sequitur or dining with an undistributed middle" [13].

One such term with which the statistical literature bristles is "interaction." The theory of the analysis of variance requires that a certain sum of components of variance shall equal the total variance. Any discrepancy statisticians call "interaction" which is simply a statement about additivity. I can tolerate this term as a device of accountancy. But many writers make of it a discrete phenomenon which exists and must be explained. I have a sympathy for the accountant who cannot balance his books and introduces an item of $27.44 which he ascribes to "miscellaneous expenses" or "petty cash disbursements." But I have no professional respect for him at all if he thinks this is really an adequate explanation of where the money actually went and sets in motion inquiries as to why so much petty cash is being spent.

Like all scientists I believe that we can make inferences from data: indeed we cannot make scientific inferences without data. And yet I am condemning those who do precisely that to the data at hand. There is an important distinction, however. It is not the business of the statistician *qua* statistician to suggest models of actual phenomena. His areas of skill and expertise, which are massive and which it is not my business to belittle, lie elsewhere.

Immense harm has been done by divorcing the analytical and experimental functions of science. This specialization might be tolerated as a concession to human weakness if in compensation there were a genuine intimate collaboration and an official policy to foster it. It would not guarantee immunity from error, but at least it would make the more crass types of error less likely. It also reduces the chance that the data of the one field will be misapplied in the other. A surmise has to meet the standards of both the empirical and the synthetic approach. The theorist may puncture the airy speculations of the experimentalist by pointing out that they are not consistent or that they are logically untestable. The experimentalist may lay to rest an elegant but unanchored formulation of the theorist. But too often neither party has attempted to establish this relationship. And the pure statistician engaged in scientific speculation is as inept and as dangerous as the scientist attempting to devise his own system of probabilistic inference. The advent of the all-purpose statisical computer program has aggravated the point almost to chaos; for fewer and fewer people understand even the logical meaning of these programs let alone their inner workings. We are coming perilously close to intellectual appeasement by incomprehension, like Euler's algebraic formula for proving the existence of God to Diderot ([7], p. 147).

While not disputing the manifest uses of regression analysis, I do not believe it is the radical solution to our problem any more than classification is.

To end, I shall try to lend substance to my topic by discussing briefly my own subject of genetics.

GENETICS

Of all the areas of biology, genetics seems at present to offer the best prospect of a *rationale* for categorization since we have an impressive body of evidence that the ultimate coinage of genetics is particulate. Yet in the history of a hundred years of formal genetics we find a titanic struggle between the classifiers and the measurers, between the approach of cardinality and that of dimensionality.

The twin founders of the field were Mendel who was almost myopically categorical, and Galton who was the champion of biological measurement. It so happened that Mendel's contemporaries were too obtuse to see the fundamental significance of his discoveries [29], even Darwin who had the greatest need of them to bolster up a very precarious theory of evolution. Thus for the first forty years Galton dominated, and laid down a very impressive theory indeed [20, 21]. When Mendel was finally recognized, his admirers made up for lost time, disregarding Galton almost completely; indeed the resulting polemics are among the most distasteful in the history of science. In 1918, Fisher in a brilliant theoretical paper [19] effected a reconciliation of the two views for those not too obdurate to appreciate it. In the subsequent quarter century the major advances in ideas were quantitative, although scientists remained loyal to the Mendelian ideas. Since the second world war we have seen a gradual erosion of the Galtonian position; and most modern geneticists have become obsessed with qualitation to the almost total exclusion of quantitation. Most now believe that the ultimate resolution of genetical problems will be categorization. The Mendelian approach has never lost its momentum. But from the Galtonian perspective, first Fisher weaned away much by showing that Galton's quantitative characteristics could be seen as the sum of large numbers of Mendelian effects. Then regulation which was first thought of quantitatively seems more and more to be Mendelian. Environmental factors were believed Galtonian; but Mendelian analysis has taken over at least some of them e.g. epistatic interaction, maternal-fetal interaction; and most recently Mendelism has started to nibble even at behavior, with what consequences it is hard to foresee. It has been becoming more and more like a one-sided struggle with almost certain triumph for the Mendelian approach and the final vindication of the category.

But then a very curious feature started to emerge. Three fields of enquiry have produced an impressive body of data in favor of, first, heterogeneity, and eventually idiosyncrasy. Increasing biological refinement has shown more and more diversity [22, 9, 14]. It is hard to continue to believe that everybody falls into a tidy number of categories. The most highly resolved system in man, the hemoglobins, has produced hundreds of different varieties [28] and the field is nothing like exhausted yet. Studies on enzymes have had a similar tale to tell. Then second the cytogenetical evidence has more and more shown what an unending chopping and changing is going on in the grouping of genes: recombination, sister-chromatid exchange and the like [8, 12, 25]. Finally studies on histocompatibility have shown such a diversity of types [1] and such bizarre statistical relationships to diseases [40] as to raise a fundamental evolutionary question. Is there selective advantage to having idiosyncrasy of tissue? If even evolutionary selection favors diversity, simplicity seems a lost cause.

Thus the categorical dream fades even as it approaches. The replacement of measurement by categorization has been successful on such a scale that the genotype is fast becoming as individual as the fingerprints. In the near future, we will doubtless conclude that everybody is biologically, as well as spiritually, unique. We shall have come full circle. This revolution, precisely because of its success, yet once again fails the fundamental desideratum of science and hence must be reckoned a blind alley.

CONCLUSION

All this discussion may sound disillusioning and even pessimistic. But this is certainly not my own state of mind. To the contrary, I find this view of knowledge exhilarating. The truth shall set us free. It is an old conceit from Pascal that the consequence of knowledge is an expanding awareness of the unknown. The more we know, the more unsolved problems we should be able to identify. Comfort is the enemy of the adventurous spirit, and too often students see what purported knowledge and truth they possess as secure and comfortable foundations to which they may regress in moments of confusion and frustration. But, in theory at least, there are no scientific beliefs I hold so securely that I would not even entertain alternative views.

We are led to reject the taxonomic approach because it involves certain assumptions which are by no means obvious, because of the practical problems it involves in application and, in some cases at least, because of its manifest inefficiency. But although the flaws of a dimensional approach are less obtrusive, they exist and in the long run may do much more harm. The only evident escape is nominalism which is in bad odor among philosophers and of course is a betrayal of the fundamental desideratum of science. It may be I am asking too much: that one cannot demand precision, economy, and detection of transcendental significance, from the same analysis. We would then have to decide on the criteria of importance. Is it most important to believe in only two human sexes, *or* to devise reliable means of distinguishing between them *or* to know what it means to be a man or a woman? But these are theoretical rather than practical problems. We can live with uncertainty provided we remain aware of our uncertainty. What I fear is that highly precise systems with spurious warrants may be built up in ignorance of these difficulties – in ethics for instance, medicine, law or philosophical psychology. I am urging caution, not preaching nihilism.

By any reasonable measure we make progress. The atomic theory of matter flourished for a century; then it became evident that the atom, the unsplitable, can indeed be split into more elementary particles, which are now legion. There is speculation that these particles are composed of more basic units still, the quarks [16]. In time we may abandon the particulate theory of matter and decide that quarks are aggregations of fundamentally amorphous matter. And that theory may in turn be replaced by a particulate one. There may be an infinite regress of hierarchies of matter. But this does not mean that seen as a scientific endeavor the whole task is futile; and by coarser pragmatic criteria, the technological uses, we have certainly derived benefit from the idea of the atom even if it has been metaphorically, as well as physically, exploded.

The fundamental, if usually implicit, controversy as to whether the phenomena of biology are categorical or continuous, as qualitative or quantitative, to be expressed in cardinality or dimensionality, may leave the philosopher profoundly dissatisfied; but again by practical criteria, the controversy, while unresolved, has had enormous heuristic value. Controversy would be desirable even if its only virtue were catalytic. But it has much wider utility. Chief among these is that it has forced us to rethink not so much what we suppose to be our basic problems as the conceptual framework in which they have been formulated. The hygienists have been saying for years that the physician thinks far too much about disease and far too little about health. Most physicians have dismissed such statements as rhetorical. Yet anyone who reads the literature or attempts to gauge the climate of opinion about such topics as antenatal diagnosis, cannot but be impressed that there is real substance in the idea of health. The law courts have provided no legal definition of the normal or diseased; and I think this shows their wisdom. On the other hand they have not hesitated [37] to give the physician powers for doing abortions for reasons concerning the nature of the fetus; but they provide no rational basis for judgment nor do they give any hint as to the principles by which they would discern whether or not this power had been abused in any particular case. Thus the academic lawyer, in developing a coherent theory of the law, has much the same problems as we do; and in certain well-defined areas such as criminal insanity, there has been extensive, but inconclusive exploration of the point [18]. The superficial might be dissatisfied; but the discerning will see that grappling with this problem the lawyer has learnt much more about the issues than if he were simply to rely on expert witnesses, a course which has been authoritatively condemned [27]. In the long run, I think that much good will come from penetrating analysis of such fundamental issues.

What are the practical implications of all this discussion?

First, there is no facile solution, universal and infinitely supple. In spite of recent social trends among adolescents, the sexes do fall into groups though not altogether unambiguously. In spite of entrenched opinion, blood pressure levels do not fall into natural groups. We must remain open to new evidence, either etiological (such as genetic mechanisms) or prognostic (such as catastrophes) which would warrant a grouping; but at present our best evidence seems to be that in diagnosis, prognosis, and treatment we should view blood pressure as a gradation. I can see much trouble for those who try to make any one approach solve all problems. Perhaps a final solution to this issue would imply that science, as distinct from technology, would cease to have further meaning.

Second, I am quite sure that perpetuation of the divorce between theory, and observation and experimentation will do more harm than good. It is deplorable that, because of narrowness of training, the investigator is robbed of that source of inspiration which lies in being able to see both viewpoints simultaneously. It is intolerable if he grows up with a supercilious attitude to, or even total ignorance of, the field he is not specializing in. For nature abhors a vacuum and the experimentor who must have a theory and has no idea that anyone has previously dealt with the issue, is apt to rediscover not the wheel but the hexagon. And in this I would fault the philosophers as much as I would the experimentalists. I must resist the temptation to preach, but we might do well to ponder the etymologies of the words "university" and "college".

Finally, I would like to express my enduring belief in the *tentative* nature of *all* our knowledge and the perils of believing that some genius of the remote past, or the present, or even in this august assembly, has laid down imperishable canons of knowledge and principles of thought. What we do as practitioners, because we must make decisions even if the decision is to postpone the decision or to do nothing, should not be confused with the subject matter of sound academic discourse. I blush to make so obvious a point to philosophers who have, if anything, fought shy of making decisions. But my trite comment is still likely to be regarded as revolutionary, even irresponsible by most of our medical colleagues. And it seems to me our business to try to change all that.

The Johns Hopkins University School of Medicine
Baltimore, Maryland

APPENDIX
THE UNCONSTRUCTABILITY OF A TEST OF THE LAPLACIAN SURMISE

The cardinality of the set of all subsets is not less than that of the set itself. Thus if the set contains n elements, in any particular subset each element may be present or absent; hence there are 2^n possible subsets. For any positive value of n whatsoever 2^n is greater than n and the discrepancy rapidly mounts as n increases.

Any analysis, in a physical medium, of a phenomenon will involve multiple particles. To write, involves arranging atoms into symbols; to speak, involves imparting to molecules of gas individual velocities and directions. But a description of all particles includes those particles which we manipulate for purposes of communication. There are thus always more particles and more relationships among them than there are messages; hence the behavior of all particles cannot be described. A strict constructionist would argue that what cannot be constructed cannot exist. A positivist position would be that what cannot exist can have no meaning. Hence systematic science is possible only if the cardinality of the descriptors is less than that of the described. The fundamental desideratum of science demands that the cardinality in complex systems be very much less.

The foregoing statement applies only to science as we understand it. The traditional view of Christian theologians – for example Aquinas [3] – is that God is omniscient, immaterial and not composite. Such a view can only mean both that God's way of knowing and God's way of thinking about what He knows are radically different from ours: that these processes are so different that we cannot call them scientific. It is not my concern here either to affirm or refute these claims of divine omniscience which involve arguments totally outside the domain of science as we understand it.

NOTES

[1] This work was supported by grant number 5P50 GMI9489 of the National Institute of Health.

[2] I do not mean this term in Lonergan's restricted mathematical sense [26], but comprehensively.

[3] There is some risk of confusion here. I mean "categories" in the modern sense of "classes". Aristotle's use of the term [4] is quite different: it is more akin to what I call below "dimensions"; and where the category is metric (e.g., "quantity"), the meanings would be identical.

BIBLIOGRAPHY

1. Albert, E.D., Mickey, M.R. and Terasaki, P.I.: 1971, 'Genetics of four new HL-A specificities in the Caucasian and Negro Populations,' *Transplant Proc.* **3**, 95–100.
2. Aquinas, T.: 1955, *Summa Contra Gentiles*, 1.11 (1).
3. Aquinas, T.: 1955, *Summa Contra Gentiles*, 1.18, 50–59, 65.
4. Aristotle: *Categories*.
5. Baker, G.A.: 1930, 'Transformations of Bimodal Distributions,' *Ann. Math. Stat.* **1**, 334–44.
6. Barr, M.L., Bertram, L.F. and Lindsay, H.A.: 1950, 'The Morphology of the Nerve Cell Nucleus According to Sex,' *Anat. Rec.* **107**, 283–292.
7. Bell, E.T.: 1937, *Men of Mathematics*, Simon and Schuster, New York.
8. Borgaonkar, D.S.: 1975, *Chromosomal Variation in Man: A Catalog of Chromosomal Variants and Anomalies*, The Johns Hopkins University Press, Baltimore.
9. Boyer, S.H.: 1972, 'Extraordinary Incidence of Electrophoretically Silent Genetic Polymorphisms,' *Nature* **239**, 453–54.
10. Bronowski, J.: 1953, *The Common Sense of Science*, Harvard University Press, Cambridge, Mass.
11. Carroll, L.: 1946, *Alice's Adventures in Wonderland*, Random House, New York.
12. Caspersson, T. and Zech, L., (eds.): 1972, *Chromosome Identification Technique and Application in Biology and Medicine*, Nobel Symposium 23, Academic Press, New York.
13. Chesterton, G.K.: 1923, *The Everlasting Man*, Dodd Mead, London.
14. Childs, B. and Der Kaloustian, V.M.: 1968, 'Genetic Heterogeneity,' *New England Journal of Medicine* **279**, 1205, 1267.
15. Collman, R.D. and Stoller, A.: 1962, 'A Survey of Mongoloid Births in Victoria, Australia 1942–1957,' *American Journal of Public Health* **52**, 813–29.
16. Editorial. Eighteenth-Century Quarks. *Scientific American* **235**, 70.
17. *Encyclopedia Americana* **3**: 1973, 618, Americana Corporation, New York.
18. Fingerette, H.: 1972, *The Meaning of Criminal Insanity*, University of California Press, Berkeley.
19. Fisher, R.A.: 1918, 'The Correlation Between Relatives on the Supposition of Mendelian Inheritance,' *Trans. Royal Society of Edinburgh* **52**, 399.
20. Galton, F.: 1889, *Natural Inheritance*, Macmillan, London.
21. Galton, F.: 1911, *Inquiries into Human Faculty and its Development*, Dent, New York.
22. Harris, H., Hopkinson, D.A. and Robson, E.B.: 1974, 'The Incidence of Rare Alleles Determining Electrophoretic Variants; Data on 43 Enzyme Loci in Man,' *Annals of Human Genetics* **37**, London, 237–253.
23. Kendal, M.G. and Stuart, A.: 1961, *The Advanced Theory of Statistics*, Vol. II, Chapter 29, Haffner, New York.
24. Lanczos, C.: 1964, *Applied Analysis*, Prentice Hall, Englewood Cliffs, N.J..
25. Latt, S.: 1976, 'Optical Studies of Metaphase Chromosome Organization,' *Annual Review of Biophysics Bio. Engl.* **5**, 1–37.
26. Lonergan, B.J.F.: 1957, *Insight*, Longmans, Green and Co., London.
27. *McDonald vs U.S.*: 1962, 312 F 2nd 847–862 (D.C. Cir.).

28. McKusick, V.A.: 1975, *Mendelian Inheritance in Man*, 4th Edition, The Johns Hopkins Press, Baltimore.
29. Mendel, G.: 1866, 'Versuche über Pflanzen-Hybriden,' *Verhdlg. Naturf. Brunn* **4**, 3–47.
30. Moore, C.B. and Weiss, J.: 1963, 'Uric Acid Metabolism and Myocardial Infarction' in *The Etiology of Myocardial Infarction*, T.N. James and J.W. Keys, (eds.), Little Brown and Co., Boston, pp. 459–478.
31. Murphy, E.A.: 1964, 'One Cause? Many Causes? The Argument from the Bimodal Distribution,' *Journal Chron. Dis.* **17**, 301–24. (Reprinted with modifications in *The Logic of Medicine* (q.v.).)
32. Murphy, E.A.: 1976, *The Logic of Medicine*, The Johns Hopkins University Press, Baltimore.
33. Popper, K.R.: 1959, *The Logic of Scientific Discovery*, Harper and Row, New York.
34. Raimi, R.A.: 1969, 'The Peculiar Distribution of First Digits,' *Scientific American* **221**, 109–29.
35. Raimi, R.A.: 1976, 'The First Digit Problem,' *American Mathematical Monthly* **83**, 521–38.
36. Ramon y Cajal, S.: 1952, *Histologie du Système Nerveux*, Consejo Superior de Investigaciones Cientificas, Madrid.
37. *Roe vs Wade*: 410 US 113 Sup Ct. Repts. 35 L Ed 2nd.
38. Rowsell, H.C., Glynn, M.F., Mustard, J.F. and Murphy, E.A.: 1967, 'The Effect of Heparin on Platelet Survival in Dogs,' *American Journal of Physiology* **213**, 915–22.
39. Sullivan, J.W.N.: 1938, *The Bases of Modern Science*, Pelican, London.
40. Svejgaard, A., Platz, P., Ryder, L.P., Nielsen, L.S. and Thomsen, M.: 1975, 'HL-A and Disease Associations,' *Transplant. Review* **22**, 3–43.
41. Thom, R.: 1975, *Structural Stability and Morphogenesis*, Benjamin, Reading, Mass.
42. Tsuang, M.T. and Tsuang, H.S.: 1972, 'A Sex Chromatin Study of Chinese School Children,' *Journal of Medical Genetics* **9**, 298–301.
43. White, P.D.: 1951, *Heart Disease*, 4th Ed., Macmillan, New York.
44. Zeeman, E.C.: 1976, 'Catastrophe Theory,' *Scientific American* **234**, 68–83.

MORTON BECKNER

COMMENTS ON MURPHY'S 'CLASSIFICATION AND ITS ALTERNATIVES'

Professor Murphy's paper is a series of reflections that converge on a complicated moral, a cautionary tale that any scientist, but especially the physicians, ought to take to heart. When he explicitly draws the moral, however, he shows a certain degree of diffidence. He warns us (p. 80) that his views admittedly sound disillusioning and even pessimistic; and that he blushes ([1], p. 82) at the necessity of making a point that must seem obvious, especially to philosophers.

His moral is indeed familiar to philosophers: that the demands of scientific inquiry (especially the construction of schemes of classification) – e.g., precision, economy, the detection of significance – are not always met in a single analysis ([1], p. 80); that neither science nor philosophy provides a royal road to the solution of the problems that exercise us ([1], p. 82); that scientists have a tendency to divorce theory and experiment ([1], p. 82); and that all our theoretical knowledge is tentative ([1], p. 82).

Let me first confess sympathy with Professor Murphy's diffidence. It is a very common feeling among philosophers. We often labor mightily, using techniques and intellectual maneuvers of the highest sophistication, only to have our colleagues and students – especially our beginning students – inform us, not always politely, that we have brought forth a mouse. In philosophy, it is impossible to conceal a fundamental fact: that we are often describing, and looking for ways to redescribe, very familiar things. Philosophers are like highly intellectualized Aesops; we tell fables about the public relations of men with overburdened donkeys and donkeys with overburdened men, and conclude with the moral that you cannot please all of the people all of the time. (We are of course likely to draw the moral in a more impressive vocabulary.) If someone answers "so what?" or "I know that already," we need not be embarrassed – at most we need to remind ourselves that in some circumstances, at some times, some people need to be reminded that you cannot fool everyone all of the time. The moral of all this: Professor Murphy should not be apologetic.

I said above that his conclusions are familiar to philosophers. This does not mean they are obvious to everyone. I disagree with a major clause, for example. I do not think all the conclusions of science are tentative – if that means that

H. T. Engelhardt, Jr., S. F. Spicker and B. Towers (eds.), Clinical Judgment: A Critical Appraisal, 87–92. *All Rights Reserved.*

we should always keep our fingers crossed, entertain real doubts about all scientific theories. I think many theories are established beyond all doubt, and that many more truths that probably don't deserve the title of theories are also established beyond all doubt. For example, we know with certainty that there are atoms and molecules; that genetic information is coded in DNA and RNA; that organic evolution took place and that Darwinian natural selection was a factor controlling its course; that the Morning Star is the Evening Star. Each of us entertains far more certainties than we do genuine doubts. The commoner view, held by Professor Murphy and many others, is, it seems to me, a legacy of the ancient tradition of scepticism, and amounts to little more (as John Wisdom once put it) than attaching the rider "possibly wrong" to each of our claims. Moreover, it also seems to me that wholesale scepticism about the sciences, so far from being the proper, modest, and sensible attitude that we ought to cultivate, is positively deleterious. It obscures an important distinction between cases in which caution is and is not warranted in the light of the facts; and it may divert philosophical attention from the task, which is of substantive value in the sciences, of describing the genuine pitfalls of inquiry.

However, this general sceptical moral is not, as I understand him, Professor Murphy's main point, so I will not pursue it further – even though it constitutes perhaps the main philosophical difference between us. His main point – again, if I understand him properly – is this: the scientist is typically pursuing a number of logically independent goals, and it is an unfortunate fact that sometimes not all those goals can be realized in a single theory. In any case, this is one of his points, and I think it is extremely interesting.

His paper has the structure: cautionary moral backed up and illustrated by philosophical reflections. I find myself in an uneasy position: I endorse the moral completely, but I have my doubts about certain of the arguments used to support it. I want to express these doubts, but in doing so add an argument or two in support of the moral.

The paper is entitled "Classification and Its Alternatives." The alternatives to classification are "clustering" and measurement in accordance with an established scale. Or, if clustering is regarded as a type of classification (as it usually is) we can distinguish between "categorization," "clustering," and "measurement," where categorization consists of classification in or out of a class according as an object does or does not possess each of a set of necessary and jointly sufficient characteristics. Professor Murphy's themes are the drawbacks and pitfalls attendant on the use of each alternative.

Let me now state my basic critical comment on the paper: I do not think there are any alternatives to classification. Thus it is a strategic mistake on Professor Murphy's part to embody his cautionary remarks in a series of suggestions that classification as such might, in some possible world, be abandoned in favor of other methods. There are no *a priori* objections to classification; nor are there any to measurement as such. Particular systems of classification, however, may be open to all sorts of criticisms: they can be inconvenient, too complex, infected in too large a degree with undecidability, useless, based on dubious or false theories, myopic, perverse, misleading, and dumb. Measures can be all of the above, and in addition invalid and unreliable. But there is no way to determine any of these faults short of close analysis of cases, in which the classification and the measure are compared closely with the things classified and measured.

I think it is easy to forget just how fundamental classification is not only in the sciences but in all intellectual activities. The availability of a yes-or-no answer to the question "Is object O a number of class K?" presupposes a system of classification; but that is no more than the supposition that the proposition "O is K" is true or false. To talk about classification is another way, slightly more complicated, perhaps, to talk about the truth-conditions of a language that contains predicates or referring expressions. If I say that Mrs. Jones has pneumonia or hypertension, I have said a bit more economically what I could say as follows: Mrs. Jones belongs to the class of those who have pneumonia and hypertension.

The only real alternative to classification, therefore, is silence; and not merely silence. Robinson Crusoe is classifying when he turns left instead of right to find his camp, or when he rejects toadstools but accepts blackberries for food. The alternative to classification is mindlessness.

From this point of view, categorization, clustering, and measurement are all special cases of classification. The dichotomies yes-no, in-out, true-false are of course inescapable, since they are trivial consequences of the class-membership relation. All the following questions are in the same bag as far as this point is concerned: Is Mrs. Jones female? Hypertensive? Exactly five feet in height? The answer in all three cases is either yes or no.

Now Professor Murphy at this point might object with some justice that I am performing precisely the kind of fast shuffle on the problems of classification that I just suggested he performed on the problem of scientific scepticism. My remarks about classification, he might say, divert philosophical attention from the important differences between categorization, clustering, and measurement, and the characteristic pitfalls of each. And (if I may continue to speak for him) all he apparently needs to do is retitle his paper, from "Classification and Its Alternatives" to "Alternatives *Within* Classification." Actually, I think this could not be done without substantial shifts in a number of his arguments; but the suggestion that attention to these differences would be profitable is certainly correct. A bit of terminology here might be helpful. Call the class of objects (in the broadest sense) that a classification system covers the "domain" of the system; call the subclasses of the domain the "pigeonholes" provided by the system; and call the properties (relational or non-relational) which are used to assign objects in the domain to pigeonholes the "diagnostic characters" for the pigeonhole. For example, the domain of a zoological taxonomic system is the class of animals; the pigeonholes are all its subclasses, including the null class and the domain itself; if, in a given taxonomic system the zoologist classifies an animal as, say, an insect by ascertaining the presence of six legs, three body segments, and a ventral nerve cord, these features are the diagnostic characters for the "insect" pigeonhole.

Professor Murphy, if I understand him correctly, associates categorization and clustering with a classification system that provides a finite or at most countable set of pigeonholes for the objects in the domain. Powerful measuring scales provide uncountable numbers of pigeonholes; an object can in principle, for example, have any length in the interval zero through one. Less powerful scales can provide fewer pigeonholes. For example, if I.Q. is measured by performance on a finite number of questions, and if age is computed to the nearest number of standard time intervals, the I.Q. scale provides only a finite number of pigeonholes. These observations are relevant because Professor Murphy seems to associate measurement with continuity and quantity; classification with discontinuity and quality; and to suggest that troubles arise when we forget, for instance, that what we regard as a qualitative difference may be only quantitative and in fact so small as to be negligible. His example is that the difference between a normotensive and hypertensive patient may be conventionally set as the difference of one millimeter of systolic pressure. And he is perfectly correct in suggesting that trouble can follow if the unwary or forgetful physician loses sight of the fact.

However, I think Professor Murphy's description of the source of such difficulties is off the mark. It is not that we are confusing quantitative and qualitative differences, and so going astray. That distinction, and his own distinction between the cardinality and dimensionality of sets, works in some cases, but even then it fails to reveal the underlying sources of confusion.

The diagnostic characters for either metric or non-metric pigeonholing may themselves be metric or non-metric or complex combinations. I doubt if the quantity-quality distinction has much future as a tool for illuminating the logic of classification systems. The important questions concern the relations between adjacent systems of classification, and the manner in which the theories *and* practical activities (I emphasize the *and*) employ the systems and their diagnostic characters. Professor Murphy's example can serve to illustrate the point. We might want to correlate hypertension with stroke, phlebitis, alcoholism, hives, etc. The point at which we draw the line between normal and high blood pressure can make a lot of difference in the correlations. But so can the diagnostic characters we choose for stroke, alcoholism, and so on. These are theoretical concerns. If the correlations are not illuminating, it is not because we have succumbed to a confusion over quantities and qualities. But these conceptions also have other than theoretical uses. A physician may say to himself "Well, only 148, hypertensive but not so bad"; the information becomes an item of clinical judgment. But he may record on the patient's chart no more than the single word "hypertension." A day later he may not remember the figure, but he may remember the word. My point here is a mild observation about philosophical method. We do not uncover the troubles about a concept like hypertension by falling back on a tried-but-not-so-true philosophical terminology, but by looking at the way the concept works in the classification scheme. Certainly, people can be misled by the term "hypertension," but the source is deterioration in information content of a message passing through channels, including (as Dr. Elstein emphasized) the processing channels of the physician's own brain. Such investigations are made to order for both philosophical and empirical study.

There is a fundamental difference, in spite of what I've been saying about their similarities, between measurement on the one hand and categorization and clustering on the other. The diagnostic procedures for metric pigeonholing must specify (at the very least) a unique ordering of the pigeonholes, and a mapping upon sequences of real numbers in their normal order. In this connection I found Professor Murphy's discussion of monotonicity puzzling. He seems to suggest (p. 70) that in the construction of measuring scales we need to be on our guard against the existence of phenomena that would produce, e.g., periodic scales, such as the sine function. He needn't worry: a function of a measure can be periodic, but a measure cannot (logically) be periodic; if the ordering and mapping relations do not preserve a unique order, then we have not succeeded in constructing a scale at all.

On the other hand, I see no fundamental difference between categorization and clustering. The former seems to me a special case of the latter. In categorization, every diagnostic character for a pigeonhole is counted as necessary, but in clustering, the decision is made to employ diagnostic characters in such a way that objects are placed in the same pigeonhole if they possess a sufficiently large number of the diagnostic characteristics, but not necessarily all of them. Thus the members of a cluster may bear no more than what Wittgenstein calls a family resemblance among themselves with respect to the specified diagnostic characters. In dealing with empirical subject-matters, the decision to count a diagnostic character as necessary for a pigeonhole is in practice revocable. There is a kind of Kuhnian conservatism, however, among users of a conceptual scheme: they tend to resist revocation unless driven to it.

I would like to close with a word of praise for clustering, and especially for the attitude of one who is willing to regard his conceptual schemes as essentially clustering systems. The attitude is a double hedge: first, against unnecessary revision of theory in the face of new empirical developments; and second – this brings me to one of Professor Murphy's main points – it is a hedge against the tendency of a classification scheme to adapt to only some of the purposes it must serve.

Pomona College
Claremont, California

BIBLIOGRAPHY

1. Murphy, E.A., 1979, 'Classification and Its Alternatives', this volume, pp. 59–85.

JOHN L. GEDYE

SIMULATING CLINICAL JUDGMENT

An Essay in Technological Psychology[1]

The world around me was . . . divided into clear-cut compartments. No neutral tints were allowed: everything was in black and white; . . . From the moment of my first stumbling words and thoughts all my experience belied this absolute position. White was only rarely totally white, and the blackness of evil was relieved by lighter touches; I saw grays and half tones everywhere. Only as soon as I tried to define their muted shades, I had to use words, and then I found myself back in a world of inflexible concepts. Whatever I beheld with my own eyes and whatever I experienced had to be fitted somehow or other into a rigid category: the myths and the stereotyped ideas prevailed over the truth: unable to pin it down, I allowed truth to dwindle into insignificance.

Simone de Beauvoir ([2], pp. 19–20)

My contribution to this Symposium is in three parts: first, I outline a 'simulation' approach to the study of judicial behavior in general; second, I give an account of an exercise in the simulation of clinical judgment; and, finally, I attempt to show how this approach to clinical argument may help to sustain relevant clinical practice in a changing world.

1. JUDICIAL BEHAVIOR

In developing this approach I will start by examining *judgment*, go on to consider *clinical* judgment, and finish with a discussion of how we might go about *simulating* clinical judgment.

1.1 Judgment

Perhaps the most relevant way to grasp the nature of the judicial process for our present purposes is through the notion of justice. Justice, personified, appears as a blindfolded figure carrying a sword in one hand and a pair of scales in the other; reminding us that judgment involves an impartial weighing of evidence, and that the outcome of the judicial process is restricted to one out of a limited number of possibilities.

On the *input* side of the judicial process we have, it seems, sensitivity to the whole range of relevant human experience; but on the *output* side, it seems, we quickly find ourselves back in a world of inflexible concepts.

H. T. Engelhardt, Jr., S. F. Spicker and B. Towers (eds.), Clinical Judgment: A Critical Appraisal, 93–113.

Does this mean, then, that to judge is to allow truth to dwindle into insignificance?

In his 'Introduction to Legal Reasoning' Edward H. Levi writes:

"It is important that the mechanism of legal reasoning should not be concealed by its pretense. The pretense is that the law is a system of known rules applied by a judge; the pretense has long been under attack. In an important sense legal rules are never clear, and, if a rule had to be clear before it could be imposed, society would be impossible." He goes on: "The basic pattern of legal reasoning is reasoning by example. It is reasoning from case to case. It is a three-step process . . . similarity is seen between cases; next the rule of law inherent in the first case is announced; then the rule of law is made applicable to the second case" ([10], pp. 1–2).

As Levi emphasizes later, the judge's initial task is to determine similarity or difference between cases, deciding when it will be "just to treat different cases as though they were the same", and so allow a rule derived from consideration of a first case to be justly applied to a second case.

To bring out the force of these points let us see how we might apply this form of reasoning to a simple example – the judgment as to whether a sample before us is to be regarded as 'black' or 'white'. We might start by assembling a collection of examples of relevant things ranging in color from black, through various shades of gray, to white. We could then arrange them in a line in front of us such that, except for the extremes, each example had a darker example on the left and a lighter example on the right. We might now set a 'lower' threshold such that it could be agreed that it would be just to regard any example to the left as 'black' and unjust to regard any example to the right as 'black', and an 'upper' threshold such that it would be just to regard any example to the right as 'white' and unjust to regard any example to the left as 'white'. To reach a judgment we could locate our sample in relation to the examples using the same ordering rule that was used to produce the original line-up of the examples, and report the result as "black", "white", or "neither 'black' nor 'white' " (which we might agree to report as "gray"), as the case may be.

It will be seen that this judicial system employs a *pair* of binary contrasts "black/not-black" and "white/not-white" to give three possible judgments. Only in the limiting situation where the set of samples chosen contains no 'grays' will it give the appearance of functioning as a two-state system; and, as we shall see later, most clinical judgment situations seem to be most appropriately represented by such a three-state system, even though their output is sometimes represented as the output of a two-state system. We note, furthermore, that this judicial system allows, by the mechanism of choosing an appropriate 'first case' and extracting a 'rule of law', for the *construction* of concepts which are appropriate to the 'second case'.

1.2 Clinical Judgment

Is the mechanism of *clinical* reasoning concealed by its pretense? If, in order to answer this question, we seek to understand this mechanism in the context of the health care system of a community, we are led to regard what I will refer to as the 'clinical encounter', as the fundamental system situation.

A clinical encounter can be thought of as any situation in which a member of the community seeking help from the system (the patient) interacts with an agent of the system (the clinician, or his or her natural (or artificial) delegate), which results in the system being presented with a problem to which its agents must respond in the light of previous experience with similar problems [8].[2]

A clinical encounter is thus an occasion for the exercise of clinical judgment, and since it is generally accepted that this utilizes a clinician's finest sensitivities, it might seem that any attempt to formalize such an activity would be a move back into a world of inflexible concepts – a step that would surely not, clinicians might feel, be in the best interests of the patient.

And yet we may remind ourselves that ideally, from the patient's point of view, the exercise of clinical judgment should culminate in a clear presentation of the patient's situation, and so prepare the way for a rational acceptance of that situation and its implications. In an important sense the patient *needs* a world of 'inflexible' concepts, provided, that is, that they are *appropriate* concepts; and, as we have seen, one way of ensuring appropriateness is to construct the concepts for the occasion.

The clinician's exercise of clinical judgment need not be regarded, then, as an indulgence in sensitivity for its own sake if it is seen as a stage in the imaginative restructuring of the situation on behalf of, and with the active participation of, the patient – in a way which allows the patient to assimilate the result of the clinician's enquiry.

Clinical arguments, and many other practically important arguments, share a feature which I shall refer to by saying that they are *hypergnostic.* They take us from the evidence presented to us (data) to conclusions (claims) that lie *beyond* the given. As Hume [9] convincingly demonstrated in his 'Treatise of Human Nature', such arguments are dubitable, since we can never be in possession of *all* the relevant evidence, but that need not deter us from using them provided we are aware of their limitations.

The nature of the gap bridged in a hypergnostic argument can be illustrated by considering an extension of the judicial system, described above, for deciding whether a sample is 'black' or 'white'.

Let us suppose that, in addition to their color our examples had been independently judged to be 'good' or 'bad', and that we had found, accepting our previous line-up, that it was possible to choose a 'lower' threshold such that all examples to the left had been judged to be 'bad' but not all examples to the right had been judged to be 'bad'; and an 'upper' threshold such that all examples to the right had been judged to be 'good' but not all examples to the left had been judged to be 'good'.

Let us now suppose that we are required to make a hypergnostic judgment as to whether a sample is 'good' or 'bad' purely on the basis of its color. As before, we locate the example in the line-up on the basis of color, but this time we employ the thresholds determined by taking into account whether the examples were 'good' or 'bad' to conceptualize 'black' and 'white'. We can now argue, hypergnostically, that a 'black' sample will be 'bad' and a 'white' sample will be good, and if we choose to do so we may decline to put forward any argument at all if the sample is neither 'black' nor 'white'. Note, though, that the terms 'black' and 'white' are now loaded with a 'good'-'bad' connotation, and their boundaries will almost certainly have shifted. This kind of change of meaning appears to be common in clinical practice – symptom words, for example, are loaded by their potential significance.

It is clear that such an argument as I have outlined for 'good' or 'bad' could produce a mistaken judgment, but insofar as following such a procedure for deriving a rule of law from a 'first case' and applying it, as a judicial *algorithm*, to a 'second case' has been found to be reliable in the past, so will it be reasonable for us to rely on it in the future, at least until something better comes along.

1.3 Simulating Clinical Judgment

Underlying the approach being presented in this paper is a view of the nature of technology which while not, as far as I can see, incompatible with more fashionable interpretations, needs to be made explicit. In a recent paper [7] I suggested that the word 'technology' could be used to denote the study of "the foundations of those human innovations on which our civilisation depends (in the attempt to discover how to account for our being able to do what we are in fact able to do, and so to set our human capacities and limitations in perspective)".

This view is designed to support the 'primacy' of technology (in this sense) with respect to other studies: it asserts, in its *boldest* form, that we can only hope to understand what we ourselves have made, and, as a consequence, that which we have *not* made only insofar as we can handle it 'as if' we had made it. The history of human thought can provide many examples of the use of contemporary technology as a source of explanatory concepts for acquiring an understanding of natural processes of various kinds, and in retrospect we often comment to the effect that it was not possible to understand a given phenomenon until such a source of explanatory concepts became available.

As stated, this view may give the impression of a rather passive attitude towards technology – human understanding has to wait until technology comes up with something appropriate to the explanatory task in hand, and so the growth of understanding is limited by the contemporary state of technology. But often, it seems to me, the need for understanding can actually stimulate the development of an appropriate technology, an observation which supports a more active view of the relationship, one which emphasizes the creative element in human understanding.

If we apply this approach to the study of human behavior we arrive at what I will call a 'technological psychology'. We seek understanding of 'human' behavior through the creation of 'humanoid' behavior, and nowadays our most powerful tool for the creation of such behavior is, of course, the computer – which we can usefully think of, at least in this context, as a general purpose *simulation* device. Since our task is to understand clinical judgment, this means that to apply this approach, we must seek to create – in the form of a computer simulation – an artificial organism that will demonstrate the powers of clinical judgment.

This is, of course, easier to say than to do and we would be foolish to overlook any opportunity for help that might come our way; indeed we would be well advised to actively seek help from any areas of human experience where such an artificial organism might be said, in some sense, to exist already. In the design of behavioral simulations our experience to date suggests that we are most likely to get help if we look to human social institutions for guidance; and so it is only natural, if our aim is to simulate judicial behavior, to turn for inspiration to a study of the judge's behavior in a court of law.

A currently emerging discipline goes under the name of 'artificial intelligence'. This is best thought of as the construction of artificial organisms capable of demonstrating intelligent behavior. It is commonly regarded as one of the latest human disciplinary innovations; but if we accept the view outlined above, we could argue that, in a sense, it started when our ancestors created the first social institution.[3]

Within the framework provided by this approach the work in which my colleagues and I are engaged is, in general, aimed at helping the clinician to make better judgments, supported by sounder arguments, on behalf of individual patients. We seek to achieve this by finding out how to create an 'artificial partner' for the individual clinician which is capable of demonstrating intelligent behavior to the extent that the clinician may come to value this partner's opinion on matters of clinical importance.

2. A SIMULATION EXERCISE

Two colleagues – Ninan T. Mathew, M.D., and Francisco M. Perez, Ph.D., kindly provided the material for this exercise. It relates to 84 patients who had undergone cerebral arteriography for a recognized clinical reason, and who, in addition, had had an investigation of regional cerebral blood flow in the minor cerebral hemisphere on the same occasion [11].

In the course of their work Mathew and Perez reviewed the diagnosis for each of the 84 patients, taking into account all the evidence available to them at the time, and for the purposes of this simulation exercise I have accepted their diagnostic categorisation of each of the 84 patients without exception. Each of the patients was placed in one of four mutually exclusive categories:

(1) Alzheimer's Disease (ALZ)
(2) Multi-infarct Dementia (MID)
(3) Vertebro-basilar Insufficiency (VBI)
(4) None of the above

The three pathological conditions include one (ALZ) where the primary lesion appears to be cellular, and two (MID and VBI) where the primary lesion appears to be vascular. It should be noted, though, that there is no *logical* reason why a patient should not fall into more than one of the first three categories, since the disorders can co-exist in the same patient. If Mathew and Perez were asked to revise their categorisation on the understanding that they could, if the clinical evidence so indicated, assign a patient to more than one of the first three categories, I think they would take advantage of this opportunity to re-classify a few of the patients; but for our present purposes we can regard the first three categories as covering, between them, the most important causes of *dementia*, that is, loss of intellectual function, in the elderly, and so we can agree to distinguish an 'abnormal' group of patients from a 'normal' group of patients, as follows:

(1) Normal – any patient *not* diagnosed as ALZ, MID, or VBI
(2) Abnormal – any patient diagnosed as ALZ, MID, or VBI

Of the 84 patients 15 were 'normal' on this definition and 69 were 'abnormal' (20 ALZ, 22 MID, and 25 VBI).

In addition to this diagnostic categorisation, regional cerebral blood flow data for 13 brain regions were available for each patient. For the purposes of this exercise we will restrict our attention to one particular kind of measurement – the rCBF10 (hereafter referred to as the rCBF), an average flow over a ten minute time interval, which is expressed in milliliters of blood/100 milliliters of cerebral tissue/minute.

For our present purposes, at least, it seems reasonable to assume that the clinicians looking after these 84 patients had grounds for believing that rCBF measurements might further their understanding of the patients' problems; and so by analogy with the hypergnostic argument which allowed us to claim 'good' or 'bad' on the basis of 'black' or 'white', it would seem that we should aim to put ourselves in a position to make a claim of 'normal' or 'abnormal' on behalf of an individual patient, on the basis of, shall we say, 'high' or 'low' rCBF.

This approach implies a rather simple set of assumptions about the likely nature of any relationship between cerebral blood flow and the kind of abnormality we are studying – but it seems, in fact, to be a view quite commonly held by clinical investigators working in the field, although there are some inconsistencies here.

It will be useful at this point to consider in more detail the form of the hypergnostic argument we hope to construct. The approach has grown out of work done some years ago [6] when, stimulated by the need to bring some order into research findings on human fatigue, I tried to develop a systematic approach to research on failures of human performance. The approach I developed came about as a direct result of reading Stephen E. Toulmin's then recently published book *The Uses of Argument* [12]. To me at that time the most important feature of this book was the claim that the practical irrelevance of most philosophical discussion of human rationality could be attributed to the fact that philosophers had settled on a highly atypical form of human argument – the so-called 'analytic argument' (in which the premises 'logically entail' the conclusions) – as the standard to which all human arguments should aspire, with the result that all but the most trivial of everyday arguments could be found wanting.

Toulmin felt that practical arguments were rather more complicated than had been assumed in the past, and suggested that one might find more relevant standard examples by studying legal arguments as used in courts of law. He further suggested that one might be able to produce a more adequate philosophy of argument by widening the scope of the approach developed by philosophers of law to include all types of practical argument. As he puts it in his introduction: "Our subject will be the *prudentia*, not simply of *jus*, but more generally of *ratio*".

This suggestion hints at the possibility of a new kind of social institution: "One may even be tempted to say that our extra-legal claims have to be justified . . . before the Court of Reason", a court that might be thought of as sitting whenever rational people gather together for discussion. This concept of a "Court of Reason" provided the ordering principle for which I was looking – to systematize our research we now needed only to think of ourselves as preparing to argue a case, for the prosecution or for the defense, as the case may be, before the Court of Reason.

In his book Toulmin went on to develop what he called a 'layout' for an argument, and while he was careful not to claim that it was in any way final, I have found that it provides a useful starting point for a consideration of clinical arguments. His general layout is as follows ([14], pp. 104):

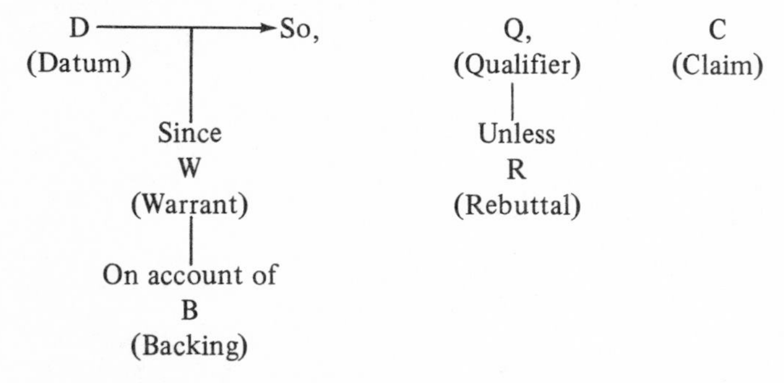

An argument allows us to pass from a datum (D) to a claim (C) this step being authorized by a warrant (W) supported by backing (B), subject to the qualification (Q) that the conclusion of the argument is not rebutted (R).

More explicitly then, we would like to be able to argue about a patient somewhat as follows: "This patient was found to have a 'low' rCBF so, presumably (unless we have reason to believe, for example, that the equipment was not working properly) this patient is almost certainly 'abnormal', since patients with a 'low' rCBF may be regarded as 'abnormal' on account of the latest evaluation of the data gathered in the Mathew and Perez study".

Having settled on this form of argument, we can now take a look at the data to see how, in practice, we might be able to go about constructing such an argument for our purposes. For convenience we will restrict our attention *at this stage* to a *single* brain region. Figure 1 shows a fairly typical situation. The segment of the vertical number line labelled "rCBF10" has been designed to allow us to represent possible rCBF values as rational numbers, and points corresponding to 'normal' and 'abnormal' values, as determined by some

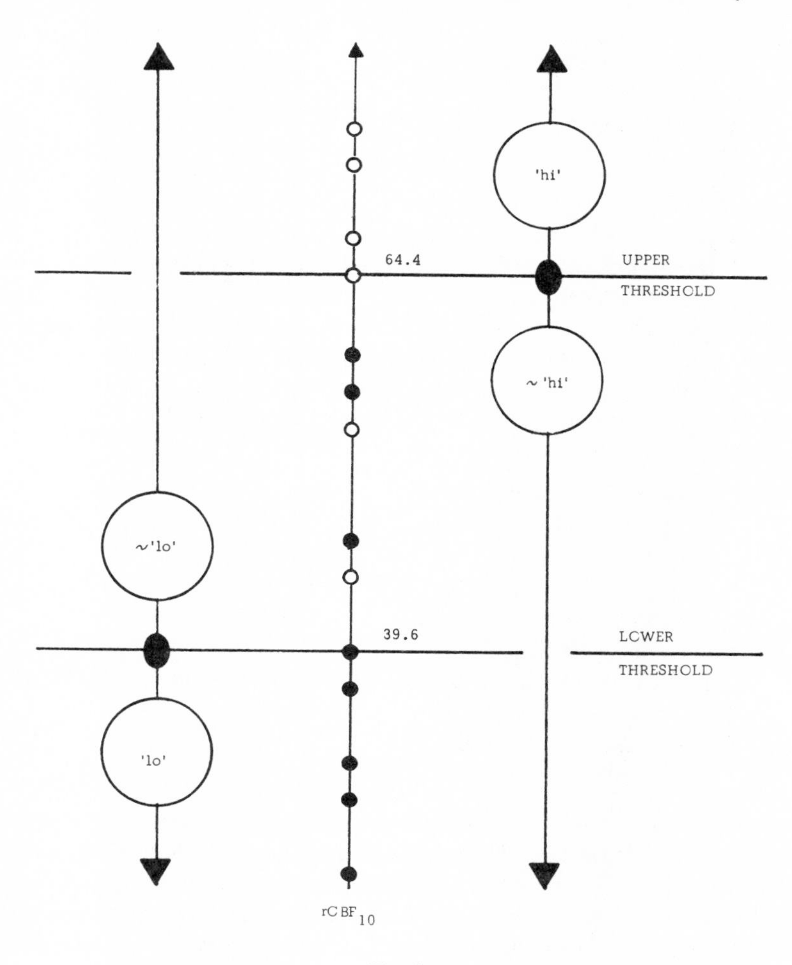

Fig. 1

typical results from the 84 patients, are represented by open (O) and closed (●) circles respectively. This is our familiar line-up and we determine 'upper' and 'lower' thresholds by finding the lowest normal value that is greater than the highest abnormal value (in this case 64.4 ml/100 ml tissue/min) and the highest abnormal value that is less than the lowest normal value (in this case 39.6 ml/100 ml tissue/min). These thresholds split the rCBF line segment into two pairs of subsegments, as follows:

(1) The upper threshold splits the rCBF line segment into an upper segment (hi) of 'high' flows, and a lower segment (~hi) of flows which are not-high.
(2) The lower threshold splits the rCBF line segment into a lower segment (lo) of 'low' flows, and a upper segment (~lo) of flows which are not-low.

It will be apparent that any rCBF value which can be represented in this way will satisfy one, and only one, of the three following conditions: hi and ~lo, ~hi and ~lo, or ~hi and lo.[4]

Let us now assume that, given the data of the 84 patients (first cases), we wish to prepare ourselves to argue on behalf of an 85th patient (second case) who meets the relevant criteria of similarity. On the basis of the above conceptualisation of the data from the 84 patient sample, we formulate a set of possible hypergnostic warrants as follow:

(1) Arguing for 'normality'
    1.1 Patients with an rCBF which is 'hi' may be assumed to be (almost) certainly 'normal'
    1.2 Patients with an rCBF which is '~hi' may *not* be assumed to be (almost) certainly 'normal'
(2) Arguing for 'abnormality'
    2.1 Patients with an rCBF which is 'lo' may be assumed to be (almost) certainly 'abnormal'
    2.2 Patients with an rCBF which is '~lo' may *not* be assumed to be (almost) certainly 'abnormal'

It will be seen that this approach partitions the possible warrants into two sets – (1.1 and 2.1), which we will refer to as 'strong', and (1.2 and 2.2), which we will refer to as 'weak'. The strong situations appear to correspond to those in everyday life in which we would claim *certain* knowledge, or, if we wished to mark the possibility of error, *almost* certain knowledge, and the weak situations to those in which we would *not* claim certain knowledge, but might, perhaps, put forward a more tentative claim, using a warrant of the form "Patients with an rCBF which is ~hi and ~lo may be assumed to be *probably* 'abnormal' " – with the rather flimsy backing (in this case) that 4 out of 7 of the examples were 'abnormal'.

In this exercise, however, our main concern is with strong arguments, and we next raise the question "How fallible are judgments supported by strong arguments?" or alternatively "How might we defend ourselves against a charge that we had been irresponsible in relying on a judicial algorithm generated by the mechanism described above?".

Let us assume that we had put forward the claim "This patient is almost certainly 'abnormal' " and that it had later been discovered that the patient was 'normal'. In response to a request to justify our original claim we could, of course, point out that, in making this claim we had deliberately used the word 'almost' to mark the possibility of such a mistake, but it would not be unfair if our opponent were to reply: "that's all very well, but how 'certain' is '*almost* certain'?"

Such a question brings out the fact, pointed out earlier, that our confidence in a judicial mechanism derives from our past experience of using it. It suggests that the only legitimate reply to the question "How 'certain' is '*almost* certain'?" would be a statement of the *form* "I have used a strong argument (of this particular strength) on p occasions in the past and on q of them I was wrong", so that we could take the ratio q/p as some indication of the likelihood of a strong claim being wrong.[5]

In order to apply this approach to our current problem we need an estimate of the likelihood of being wrong when we make a strong claim with regard to an 85th patient. To make this estimate we fall back on an appropriate precedent, the closest to hand being our experience of making such a claim with regard to the 84th patient (second case) on the basis of our knowledge of the preceding 83 patients (first case(s)).

In a real life clinical situation, of course, such a series of 84 patients would have presented to the clinician in a particular order, this being determined by many factors ([15], pp. 173–208). However, provided that the criteria of similarity for the particular problem we are studying are met, it would seem to be in the best interests of the nth patient to treat each of the preceding n-1 patients as of equal potential relevance and to disregard the order in which they in fact presented, until, at least, we have a good reason for doing otherwise.

We will refer, in this context, to the act of crossing the boundary between the known and the unknown as 'bridging the nth/n+1th patient knowledge gap'. With no unique 84th patient we can think of each patient occupying this position in turn, and base our estimate of being wrong when judging the 85th patient (on the basis of the preceding 84), on our experience of how often we were wrong when, on 84 occasions, we judged each of the patients on the basis of our knowledge of the preceding 83 patients. Our confidence in our judicial mechanism thus derives not just from a study of a particular set of patients in the order in which they presented in real life, but from what we might call a judicial *rehearsal*, involving a set of possible situations chosen to be appropriate precedents for our problem situation.

We can now resume our discussion of the matter of "how 'certain' is '*almost* certain'?". If we are to be consistent with the use of the word 'certain' in the language of common discourse, we should, it seems to me, insist on a *zero* error rate at judicial rehearsal, and derive our estimate of what we mean by '*almost* certain' from the observed error rate when we actually use such arguments in practice.

When this approach was applied to the task of bridging the 83rd/84th patient knowledge gap of the Mathew and Perez material, it was found that it was not possible to find a *single* brain region such that a strong argument based on data from that brain region alone would be substantiated on more than 54/55 occasions, an error rate of 1.8%. It is of interest to note that the median error rate for the 13 brain regions, used singly in this way, was about 4% – not so far from the statistician's '5% significance level' first recommended by R.A. Fisher as a reasonable one for practical purposes.[6]

But given this rehearsal data for the task of bridging the 83rd/84th patient knowledge gap, we can formulate a second order rule by taking, for example, the best region with an error rate of 1.8%, together with the next best region with an error rate of 2.0%, and stipulating that we will only put forward a strong claim if *both* regions yield – under first order rules – conceptualized data supporting consistent strong arguments for one of the alternatives. Using the best two regions in this way it was found that we could put forward strong claims in favor of abnormality in 43/84 (51%) of cases.

We can go on to broaden the base of our claims by extending this procedure in such a way as to increase the number of strong claims that can be put forward at rehearsal of the task of bridging the 83rd/84th patient knowledge gap. For example: a judicial algorithm was formed by recruiting 4 brain regions, and using a second order rule of the form "if the data from any 2, or more, out of the 4 regions allow consistent strong arguments – under first order rules – for one of the alternatives, put forward a strong claim for that alternative, otherwise do not". Using this judicial algorithm we find that strong claims in favor of abnormality could be put forward in 56/84 cases (66%). This result is illustrated in Figure 2.

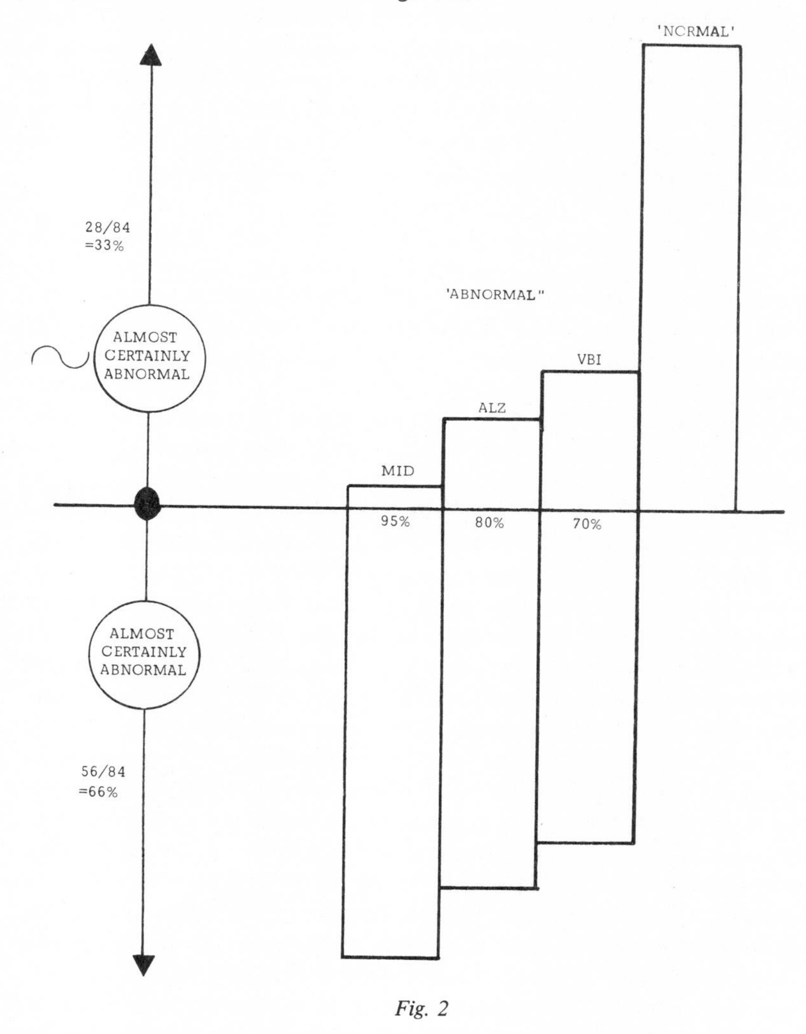

Fig. 2

We see, therefore, how study of judicial performance in an appropriate rehearsal situation can lead to the development of second order judicial rules which use information efficiently and at the same time allow stringent criteria of certainty to be applied to the judgments based on them. These second order rules will, of course, themselves need to be evaluated and in our search for an appropriate precedent we can, of course, go one stage further back. And so it is that we find ourselves using data from a rehearsal of bridging the 82nd/83rd patient gap to evolve second order rules applicable to a rehearsal of bridging the 83rd/84th patient gap, so that we can estimate the reliability of a judgment that seeks to bridge the 84th/85th patient gap. Is it going too far to see this as a simulation of certain aspects of self-conscious behavior?.

3. CLINICAL JUDGMENT IN A CHANGING WORLD

Stephen E. Toulmin, in his book *Human Understanding* ([13] – the first volume of which appeared in 1972), develops an evolutionary approach to knowledge in terms of a 'populational' account of conceptual change in intellectual disciplines. From the standpoint of Toulmin's approach, "men demonstrate their rationality not by ordering their concepts and beliefs in tidy formal structures, but by their preparedness to respond to novel situations with open minds – acknowledging the shortcomings of their former procedures and moving beyond them." The emphasis in his approach to rationality is thus on 'change', on the circumstances under which and the means by which men change their concepts and beliefs.[7]

We can distinguish between short-term and long-term mechanisms of adjustment to changing circumstances. Imagine a situation in which a tribe, not color blind, lives in a world of varying shades of gray – like a black and white movie, and whose language contains only these two color words (as in the Jalé language of the New Guinea highlands). Imagine now that this tribe is suddenly required to take account of colored objects in its everyday life. The adaptation will not be possible within the language in its present form, it will require the introduction of color concepts, and a corresponding 'ascent' of the scale of color terminologies towards the situation in English with its eleven primary color terms [3]. This linguistic development, in addition to introducing new color terms, will result in a redrawing of the boundaries of 'black' and 'white', and will allow a whole new range of discrimination to be handled by members of the tribe in their everyday life [8].

In a very crude way this imaginary example gives us some insight into what seems to happen when a powerful new information source is introduced into clinical practice, a situation which is well illustrated by the recent arrival of computerized tomography (CT scanning for short). The impact of this invention on clinical practice has, so far, been greatest in neurology, where it has given the clinician access to a potentially bewildering profusion of high resolution pictures of cross-sections of the living human brain (but not those pictured in the traditional textbooks of neuroanatomy), each available in a virtually inexhaustible variety of false colors.

The clinician's professional language painstakingly developed over the years to deal as well as possible with the clinical problems presented to him on the basis of much poorer and less direct sources of information than the CT scan, lacks the conceptual equipment to take full advantage of the new situation.

The only general solution to this problem, short of abandoning the whole enterprise of seeking to provide ourselves with richer sources of information, seems to me to be to go one stage further, and to seek to use our information handling technology to facilitate the development of new concepts, or, to employ the language of 'artificial intelligence', to learn how to employ an artificial partner to acquire these concepts for us [8].

In our simulation exercise we were responding to a suggestion, inherent in Toulmin's aproach, to the effect that if we wish to study *rational* behavior we should focus on what happens when we try to bridge the gap between the known and the unknown.[8] We used, deliberately, a situation in which our 'artificial partner' – in the form of an exteriorization of our own thought processes – was faced with the problem of learning how to use data from a source to cross that gap – to make a hypergnostic inference; and as such it illustrated the way in which we might tackle the problems posed by CT scan data (a task which is, incidentally, one of the main interests of our research group).

However, it may be objected that, even in these days of CT scanning and the like, such situations are not typical of clinical encounters, that the exercise of clinical judgment does not normally involve the establishment of new knowledge, but rather the application of well-established knowledge to new situations; that is, it is concerned with warrant-using rather than warrant-establishing activities. But this may be to conceal the mechanism by the pretense, a danger greater in clinical than in legal practice because of the relatively private nature of the clinical as opposed to the legal encounter.

It is at this point that the discipline I referred to earlier as 'technological psychology' becomes pertinent. In using the relevant behavior of appropriate social institutions as inspirational sources for the creation of computer simulations of judicial behavior, we are equipping ourselves with a means of understanding an individual's judicial behavior; just as the invention and use of suitable chronoscopes by 18th century astronomers allowed them to begin to understand temporal aspects of individual human decision-making behavior – by enabling them to begin asking questions about that behavior that could not even have been posed without an appropriate, already existing, 'chronological' discipline.[9]

One obvious reason why the processes of law are more transparent to us than clinical processes is that courts of law, serving as they do a higher level societal function than the clinician's office, have a much lower rate of turnover of cases and, as a consequence, it has for a long time been possible for them to insist on much higher standards of record keeping than have been feasible in regular clinical practice until relatively recently.

The computer, by enhancing the clinician's ability to monitor his own activity in the circumstances of the high rate of turnover of the clinical case, may allow more explicit application of the judicial processes outlined earlier to clinical practice, thus enabling him to commit himself more wholeheartedly to a continual activity of preparation for the next case, an ongoing rehearsal entered into in the knowledge that the curtain may at any moment rise, unannounced, on a live performance.

Nevertheless, some of the implications of this approach may become clearer if we consider how we might apply it in a relatively familiar situation. Let us imagine that we ask our colleagues Mathew and Perez to list all the sources of data that they feel they *might* have used in reaching the diagnostic categorisation which we accepted, without question, for the purposes of our simulation exercise. Let us further imagine that we have been able to acquire and appropriately represent data from each of these designated sources, for each of the 84 patients.

We are now in a position to apply our approach, suitably modified to accommodate the various new types of data, to the construction of a set of arguments in just the same sort of way that we did with the rCBF data in the simulation exercise. The resulting judicial algorithm would not pretend to be a representation of the mental processes of either Mathew or Perez, but it would, within its limitations, be a demonstrably effective means of reaching clinical judgments and as such something that could be adopted by the clinical community as an agreed basis for the categorisation of patients. It would seem to be in some such way as this that we could hope to arrive at generally accepted criteria of diagnostic categorisation – as has already been achieved, for example, for some cardiac disorders [5], and which is an important early step in clinical research – what is or is not 'abnormal' is *not* for the public to decide.[10]

The establishment of such generally accepted judicial algorithms would open the way to further developments, as exemplified by our exercise of seeking to incorporate a new source – rCBF measurement – within the framework of clinical understanding provided by the information derived from traditional measures. Such incorporation might lead to proposals for revision of diagnostic criteria, the most drastic of which might be that we would agree henceforth to define 'abnormal' in this context as having a 'positive rCBF test'. That is, we would introduce a new concept constructed to have the property that '∼positive' has, by definition, the same meaning as 'negative', and '∼negative' has the same meaning as 'positive'.

A somewhat less drastic proposal would be to substitute the rCBF test for one or more of the traditional sources – perhaps on the grounds that it would reduce the cost of the decision. Such proposals are currently pertinent to the clinical justification of expensive procedures like CT scanning.

Finally, I would like to raise some general questions concerning the matter of claiming 'certainty' for clinical arguments. In my view one of the most important aspects of the approach I have outlined relates to the distinction between strong and weak arguments. This distinction allows us to claim (almost) certain, 'black and white', knowledge about *some* things, but requires us to refrain from such claims with respect to other things. Rather than requiring us to lump everything together and, as it were, smear a layer of uncertainty over them all, it allows us to partition our knowledge into what we do, and what we do not, claim to know something about.

As far as clinical judgment is concerned I can see no other way to proceed if, in the clinical encounter, we are to do the best we can for the individual patient. As a rule, lack of certainty should be a spur to seek new evidence and to revise our judicial algorithms, not the occasion for a tentative pronouncement; only when we have no alternative and must decide in the absence of certain knowledge, should we accept having to fall back into the 'gray' world of weak, probabilistic, arguments.

But what justification do we have for the hypergnostic leap across the gap between the unknown and the known? In one sense this question is answered when we make explicit the backing for the inference warrant used but we still feel entitled to ask such questions as "What principles underly this?", "How do we account for the fact that it works?"

We might try to answer these questions by saying that if the descriptions of two patients are found to be, in some sense, *near* to each other we hypergnose that their behavior will be, in some sense, *similar*, because we have constructed our definition of nearness to be such that our experience will support such a conclusion. The concept formation procedure described earlier is a means, admittedly crude, for constructing a conceptual system with the requisite properties. But under what circumstances will such an exercise be successful? A problem posed by Banerji [1] is pertinent to this question. In (a slight modification of) his situation our artificial partner is required to evolve a judicial algorithm to hypergnose the next patient – 72(?) – in the following series 11(a) 32(n) 57(a) 74(n) 26(n) 43(a) 58(n) 85(a) 99(a) 20(n), where the two digit numeral represents the clinical data and 'a' denotes abnormal and 'n' normal.

Banerji asks "how sure are we that 72 is normal?", and his 'answer' is to the effect that if our rule is "we hypergnose the patient as normal if the numeral, interpreted as a natural number, is *even* and abnormal if it is *odd*", then we are fairly confident – by commonly accepted criteria – otherwise, we are not. It is clear that the judicial mechanism used in our simulation study would not arrive at this solution unless we had presented the data in the form of a base 2 representation of the numbers, with instructions to ignore place value – in which case our partner's judicial mechanism would learn to ignore everything except the units column.[11]

This example shows that the solubility of the hypergnostic problem may depend on having, or finding, an appropriate representation of the data, appropriate in the sense that it manifests pertinent criteria of nearness. Our knowledge of cerebral blood flow, for example, gives us no reason to suspect that the parity of such natural numbers as we might use to represent flow values will be relevant to the task of hypergnosing the normality or abnormality of the patients from whom the data comes. If it were it would imply that all flows represented by even numbers were in some sense near to each other and distant from all flows represented by odd numbers, even though they alternate on the natural number line.

On the other hand our knowledge of computer scientists and their activities is such that we are not unduly surprised if we find that they have invented a game in which such a property *is* regarded as relevant.

The issues brought out by Banerji's problem are pertinent to the question, dominant in the minds of many workers in the field of artificial intelligence, of how we educate an artificial organism about the 'real world'. Our 'solution' to this problem, as will be apparent from the general nature of the approach outlined in this paper, is to teach our artificial partner with carefully chosen examples, using appropriate representations of the relevant data – so that our partner acquires knowledge of the real world through the *data*, more specifically by discovering which data representations lead to fruitful criteria of nearness, and so to useful judicial algorithms; in other words by the exercise of scientific thought.

Remember, then, that [scientific thought] is the guide of action; that the truth which it arrives at is not that which we can ideally contemplate without error, but that which we may act upon without fear; and you cannot fail to see that scientific thought is not an accompaniment or condition of human progress, but human progress itself. W. K. Clifford [4]

Baylor College of Medicine
Texas Medical Center
Houston, Texas

NOTES

* I would like to acknowledge the help of my research assistant, John B. Zuckerman, who has been responsible for the implementation of the program ZAP, some aspects of which are discussed in this paper. Paul K. Blackwell and John B. Zuckerman were kind enough to give me the benefit of their comments on an earlier draft of this paper, but I would not wish them to be held responsible for any of the views expressed here. The work was supported in part by NIH-NS-09287.

[1] The text of the paper as printed here has been slightly modified from that submitted prior to the Symposium for Professor McMullin's commentary. It incorporates both suggestions made by colleagues and by Professor McMullin himself. Where I have felt it to be inappropriate to accommodate a suggestion by changing the original text I have added a note to deal with the point raised. Issues not covered in either way are dealt with in my rejoinder: 'A Reply to Ernan McMullin'.

[2] I introduce the 'system' approach here in order to draw attention to a self-imposed fundamental requirement of the work in which I am engaged – that its results should be applicable to the task of understanding a community health system. The suggestion is that the judicial skills which result from clinical experience are not to be thought of as being embodied *just* in the individual clinician, but also in the social institution in which he or she works.

[3] Throughout my experience of designing computer-based systems, I have been frequently struck by the way in which successful designs seem to evolve from attempts to copy an appropriate precedent from the everyday world. A computer-based filing system, for example, can be modelled on a clerk's interaction with a traditional filing system in, say, a business office. The resulting, computer-based, system will 'go beyond' the original for a number of reasons, in particular because it is free from many of the constraints operating in the original environment.

This kind of experience leads to the suggestion that if we want, say, to create an intelligent artificial organism, we should try to copy not the behavior of an individual considered in isolation, but in the context of a relevant social institution; and that we should look to the removal of constraints inherent in the original situation as a powerful source of new potentialities.

[4] This procedure specifies the concepts 'hi' and 'lo' in terms of well-defined sets and their complements.

[5] My purpose here is just to indicate the general lines on which one might respond to such interrogation. There would seem to be a number of ways in which one could, for example, *rank* strong claims according to the strength of the backing.

[6] If the 'normal' and 'abnormal' distributions of rCBF values for any single brain region overlap, there will be *up to two* mistaken strong claims, if we just argue in accordance with a first order rule – as McMullin quite rightly points out in his commentary on my paper. This is the reason, as I point out in my rejoinder, that we go on to use data from several regions and argue in accordance with a second order rule. I deal with these points further on page 141 of 'A reply to Ernan McMullin'.

[7] Here I am acknowledging my debt to the writings of Stephen Toulmin, which I have found to be a valuable source of inspiration. I am well aware that Toulmin's use of the word 'concept' may not correspond to my own, which is closer to that used by linguists, for example, in work on color concepts. However, it would not surprise me if pursuit of the approach I have outlined led to the emergence of 'higher' concepts. We already see the beginnings of this in the situation in which we introduce a second order rule in order to allow us to put forward a strong claim. This second order rule involves a concept embodying a precise notion of 'family resemblance' (see Note 6).

[8] Again I am well aware that Toulmin might be more concerned with change at a 'higher' level, but I nevertheless feel that his emphasis on studying what happens when we try to bridge the gap between the known and the unknown is, at the very least, of considerable heuristic value at the 'lower' level at which I am working.

[9] I am referring here to the fact that it was not until precise measurement of times became important in astronomy that it was discovered that it 'takes time' for a human being to make a decision. I find that there is little general awareness of the fact that it is only for the last 150 years or so that it has been accepted that the speed of human thought is not infinite. We are still a long way from understanding all the implications of the finite speed of thought in – for example – the design of clinical systems, but at least we have a 'chronological discipline' from which to start. When it comes to understanding human judgment, it seems to me, we are further behind because we do not yet have a relevant 'technological' discipline to apply to the problem.

[10] The allusion here is to a remark of Wittgenstein: 'what is or is not a cow is for the public to decide'. It is quoted by Stephen Toulmin in his *The Philosophy of Science* ([14], p. 51).

[11] Some people have reacted to the 'odd-even' distinction by saying that 'anyone' could spot it. It is, perhaps, not too difficult to do so if you are already familiar with the distinction between 'odd' and 'even', but our 'artificial partner' is not familiar with this distinction until we provide the requisite educational experience. We can do this by setting a hypergnostic task with data represented in such a way that the relevant attribute is available for selection. For further discussion of this very interesting problem I refer the reader to Banerji's original paper.

[12] In retrospect it occurs to me that my reasons for approaching the subject of clinical judgment in the way that I did may not be as apparent to the reader as I assumed they would be at the time I wrote the paper.

My main concern at that time was to try to provide concrete examples of the *kinds* of problem that I was facing in my research on computer-based aids to clinical decision making, and to do so in such a way as to bring out the related philosophical issues.

The crude procedure that I described for conceptualizing high and low blood flow was in fact the starting point of my work on the problem of how to encode clinical data for input to an 'artificial partner'. I had already gone some way beyond this when I wrote the paper, but to have pursued the matter in greater detail would have been to run an unjustified risk of unbalancing the presentation.

It may, however, be of interest to readers to know that the latest version of the program ZAP conceptualized the situation in such a way (see page 109) that 16 patients out of the 84 have a 'negative' 'rCBF test' (14 normal and 2 abnormal) and 68 have a 'positive' 'rCBF test' (1 normal and 67 abnormal). If, in other words, we continue to accept the Mathew and Perez classification, the applicability of the investigation has been widened at the cost of a loss of ability to put forward a 'strong' argument on the basis of the 'rCBF test' alone; additional information is needed, but probably not much.

BIBLIOGRAPHY

1. Banerji, R. B.: 1973, 'Simplicity of Concepts, Training and the Real World', in A. Elithorn and D. Jones (eds.), *Artificial and Human Thinking*, Elsevier Scientific Publishing Company, Amsterdam, pp. 70–79.
2. de Beauvior, S.: 1959, *Memoirs of a Dutiful Daughter* (transl. by J. Kirkup), The World Publishing Company, Cleveland.
3. Berlin, B. And Kay, P.: 1969, *Basic Color Terms*, University of California Press, Berkeley.
4. Clifford, W. K.: 1872, 'On the Aims and Instruments of Scientific Thought' in L. Stephen and F. Pollock (eds.), *Lectures and Essays*, Macmillan and Company, London, 1879. Quoted as frontispiece to: Dover Publications, Inc. edition of *The Common Sense of the Exact Sciences*, 1955.
5. Feinstein, A. R.: 1967, *Clinical Judgment*, Williams and Wilkins, Baltimore.
6. Gedye, J. L.: 1964, 'Transient Changes in the Ability to Reproduce a Sequential Operation Following Rapid Decompression', Institute of Aviation Medicine Report No. 271, Farnborough.
7. Gedye, J. L.: 1972, 'A Computer-based Aid to Self-Awareness', *Proceedings of British Institute of Management Conference '72 on Computers and Stress in People*, pp. 5–6.
8. Gedye, J. L.: 1977, 'A Technological Approach to the Problems of Contemporary Health Care Systems', in *The American Biomedical Network – Health Care Systems in America: Present and Future*, S. B. Day, R. V. Cuddihy and H. H. Fudenberg (eds.), Scripta Medica & Technica, South Orange.
9. Hume, D.: 1739, *A Treatise of Human Nature*, Everyman Edition, J. M. Dent, London, 1911.
10. Levi, E. H.: 1948, *An Introduction to Legal Reasoning*, The University of Chicago Press, Chicago.
11. Perez, F. I., Mathew, N. T., Stump, D. A., Meyer, J. S.: 1977, 'Regional Cerebral Blood Statistical Patterns and Psychological Performance in Multiinfarct Dementia and Alzheimer's Disease', *Le Journal Canadien Des Sciences Neurologiques*, 53–62.
12. Toulmin, S. E.: 1958, *The Uses of Argument*, Cambridge University Press, London.
13. Toulmin, S. E.: 1972, *Human Understanding*, Vol. 1, Clarendon Press, Oxford, England.
14. Toulmin, S. E.: 1968, *The Philosophy of Science*, Arrow Books, London.

ERNAN McMULLIN

A CLINICIAN'S QUEST FOR CERTAINTY

John Gedye began his paper by quoting Simone de Beauvoir who laments the fact that words force her into rigid categories, into blacks and whites that ill match a world in shades of grey. Inexorable, then, she cries, that in such a world truth should dwindle into insignificance. This is a cryptic saying to introduce a paper which will seek to strip the clinician of his customary grey language and force him to "black and white" assertions (p. 109) and "stringent criteria of certainty" (pp. 105–106). "Only when we have no alternative. . .", Gedye admonishes his fellow-clinicians "should we accept having to fall back into the 'gray' world of weak probabilistic arguments" (p. 109). Are we to infer, then, that in the world of the clinician truth is doomed to dwindle into insignificance? I cannot believe that this is what Gedye intended, yet this *is* the moral of de Beauvoir's remark.

Even more sombre is the quotation with which the paper ends. The mathematician of a century ago, W.K. Clifford, tells us that "scientific thought is not an accompaniment or condition of human progress, but human progress itself" [3]. Only a 19th century man could seriously propose that scientific thought is synonymous with human progress. True, there have been people born in the twentieth century, like the late Jacob Bronowski, for instance, who stubbornly maintain the Enlightenment faith that the growth of science is the true measure of the ascent of man. But they have to ignore or explain away much of our sad century, and indeed much else besides. I find Clifford's sentiment rather troubling in a paper advocating an increased reliance on machines, or as Gedye terms them, "artificial partners", in clinical work. Are we to infer that the future improvement of the clinical encounter between doctor and patient depends entirely on the extent to which the scientific rationality of that encounter can be sharpened? I surely hope not, though this would be the implication of Clifford's position. I suspect that Dr. Gedye intends only the weaker suggestion that the quality of the clinical encounter could be appreciably improved by a more intelligent use of scientific method. With this, I would, of course, fully agree. But the problem here will be to discover just what this recommendation comes to. 'Scientific method' is at best a weasel phrase. At worst, it can be a trumpet calling troops into the wrong battles.

H. T. Engelhardt, Jr., S. F. Spicker and B. Towers (eds.), Clinical Judgment: A Critical Appraisal, 115–129.

Let me recall just one of those battles, and its most illustrious general. Francis Bacon, in the preface to his *New Organon* (1620), promised to "lay out a new and certain path for the mind to proceed in". He was contemptuous of the traditional Aristotelian science which had tried, futilely in his view, to come to a knowledge of nature by "the naked force of the understanding" alone. If, he reminds us, the engineers and artisans of the ancient world had similarly tried, without the aid of technology, to move great obelisks, they would never have succeeded. Yet succeed they did, and because they called on the help of instruments and machinery. The scientist must do the same, except that here his aids must take the form of a *method*, a set of rules that anyone can learn. With such a method, science can become, like technology, a progressive enterprise, ever marching to new triumphs. What is needed, in short, is that:

> the mind itself be from the very outset not left to take its own course, but guided at every step, and the business be done as if by machinery. [1]

And he believes he has discovered such a method.

This Baconian program seems close to what Gedye has in mind. Its two characteristic assumptions are that the scientist has available to him a well-defined method, a series of steps which if followed will necessarily lead to the truth, and second, that science is not a matter of probability or approximation but of true causes known with certainty. If the method but be faithfully followed, certainty can (Bacon insists) eventually be attained. His method works "as *if* by machinery"; he does not think of relying directly on machinery, indeed discounts the use of instrumentation to extend the range of the senses. The method of induction works as well as a machine would, he thinks. What need of further help? Yet it is an easy extension of his program to see in technology not only an analogue but a direct aid, and to recognize that a formalizable method is one which can be programmed into a computer.

Bacon's stress on method and on science as certain and necessary truth were not peculiar to him. They were found, to one degree or another, in virtually every major theorist of science of his century. But it was in his works that it found its most durable expression. There are echoes of this program in John Gedye's paper. There is the same uneasiness with the "almost certain", where the *certain* is what should be sought. There is the same conviction that an orderly inductive method will do the trick, where the naked force of the understanding (read: of the clinician's poorly organized intuitive skills) has failed.

My comment will be divided into three parts. First, I will try to extract a most important general point that I take Gedye to be making. It is one with which I fully agree, one which indeed I take to be uncontroversial, but one which, unfortunately, has not been enough stressed in the past. Second, I will deal with two analogies that he constantly alludes to, analogies which I believe to be unhelpful, and indeed unnecessary, to his case. Third, and principally, I will touch briefly on some of the difficulties in the general proposal he makes, and suggest in passing a rather different analysis of clinical judgment.

1. THE CLINICIAN AS EXPERIMENTER

Alvan Feinstein declares that his work, *Clinical Judgment*, (1967) is:

> intended to propose clinical studies not as a preferable alternative to laboratory research, but rather as an additional and equally valid opportunity for scientific clinical investigation. ([4], p. 38)

A century ago, it would not have been possible to include the work of the clinician in experimental medicine proper. But enormous advances in laboratory research have changed this.

> These improvements have removed two insurmountable handicaps that prevented Claude Bernard and older generations from seeking science in the active experiments of therapy. A patient's disease can now be reproducibly identified; the agents of therapy can now be isolated and effective. Consequently, in every act of contemporary prognosis and therapy, clinicians can perform active scientific experiments as deliberate and distinct as any work done in the laboratory. ([4], p. 39)

Feinstein goes on to complain, however, that clinicians remain deplorably uncertain of when and how to turn their work to experimental purpose; they fail, he charges, "to exploit their opportunities for achieving scientific performance in the routine experiments of the bedside" ([4], p. 39). I take it that Gedye is indicating one fruitful way in which these opportunities could be exploited, namely the way in which Mathew and Perez did so in their study of the correlation between the rCBF in various regions of the brain and specific forms of dementia.

What is required here is that there should be a careful tabulation of cases, with a reporting of all factors thought to be relevant as well as of the diagnosis reached. Such tabulation requires not only an extra degree of patience and precision on the part of the clinician but also a convenient means of testing for correlations. It is here that the advantages of the computer become obvious. Not only does it have virtually unlimited storage, but it also allows rapid organization and precise scanning of the data. It is the sort of device that Francis Bacon would surely have greeted (even set out to build), had he more fully realized the complexity of the task his scientists faced.

Needless to say, this sort of tabulation of clinical data and this sort of use of the computer are by now quite common. It is the staple method of the epidemiologist, for example. Feinstein's point is that it is not used *enough*, that there is an immense potential resource in clinical examinations which could extend medical science appreciably were it to be more systematically tapped. Gedye's emphasis is not so much on what the clinician can do for science, as on what the clinician's science can do for the clinician. What is of concern to him is not a correlation between rCBF and dementia in a group of 84 patients, or a generalization of this to a claim about dementia generally, but rather what the correlation established over the first 84 patients allows him to tell the 85th. His is the clinician's anxiety over what to tell the individual patient, over the degree of confidence he can properly exhibit in regard to his diagnosis.

Of course, one cannot really separate the scientist and the clinician here. If the scientist comes up with a statistical correlation between a particular trait or set of traits and a particular disease, the clinician can assuredly *use* this correlation to tell the next patient he discovers with this condition what his chances are. This is the sort of thing that clinicians do all the time. They may derive their estimates from published data or from their own experience or a combination of the two. It is Gedye's point that if they are to make use of their own experience to the full, it should be rendered as far as possible into a state where statistical analysis can replace vague estimates based on memory.

His point is, however, not *just* this, because he wants to construct a method of analysis which will allow the clinician to speak in terms of *certainty* and not just of probability. It is this which is distinctive about his proposal; it raises, I think, a whole host of difficulties. But before we come to those, I want to refer briefly to two analogies that pervade the language of his paper.

2. THE LEGAL ANALOGY

The first of these is the analogy between the clinician and the judge, between the clinic and the law-court. Following Stephen Toulmin John Gedye believes that each of these exemplifies what he calls the "judicial process", and that furthermore "the most relevant way to grasp the nature of the judicial process for our present purposes is through the notion of justice" (p. 93). If the clinician wishes to "simulate judicial behavior" then, he should "turn for inspiration to a study of the judge's behavior in a court of law" (p. 97).

My first remark on this is that it is surely to explain the obscure by the more obscure. But my objection runs much deeper than that. The disanalogies between legal and clinical reasoning are so many and so basic that this move, to my mind, is more likely to confuse than to illuminate. Let me mention some of the more obvious disanalogies. The judge determines fact somewhat as a historian might; the action he is judging is ordinarily a past one. There is no sort of reproducibility, no way to "run it through again". Whereas the clinician can ask for further tests when ambiguities arise. Second, the law the judge is applying is usually a written one; his task is to interpret it, to ask what it means, in order to decide whether it applies to this case. The clinician is applying empirical generalizations, based on repeated observation and testable in individual cases such as that of the patient before him. The outcome of the clinician's work serves as a direct test of the adequacy of the generalizations he is working with. There is no similar constraint on the work of the judge. Third, the clinician can predict what will happen, if his diagnosis is correct. The judge is not in any way involved in prediction. Fourth, the judge's decision constitutes of itself a precedent which may serve as a warrant for later decisions. There is nothing analogous to this in medicine.

But rather than multiply these disanalogies, it may suffice to note the basic differences of function between the two. The judge's aim is to see that justice is served. Not truth, *justice*. In the law-court, various kinds of evidence are disallowed, even though they would help in the search for the truth of the matter, because they are held to compromise justice. Furthermore, when the judge passes judgment, he passes it *on* someone, and it is effective in bringing about a state of affairs (e.g. conviction, punishment). To *pass* judgment is very different from *making* a judgment as a clinician does, or as a golfer might, in choosing the right club for his next shot. The word 'judgment' is radically ambiguous here, and it is this ambiguity that has led to the confusion.

Among philosophers of science, Toulmin is the one who has urged the legal analogy as a means of understanding the work of science. His argument has been widely debated. I do not have the space here to criticize it in detail (see [5]). His notion of "rule" or "inference-ticket", derived from Wittgenstein, leads him to represent scientific laws, not as assertions about the world, but as rules which allow one to argue from one case to another similar case. And, of course, the judge in the law-court is apparently doing the same thing, using rules to determine whether the case before him is, or is not, similar to a particular previous case. I will leave aside how well this describes the judge's task.

What I want to challenge is the positivist supposition that scientific laws, and this includes well-established empirical correlations used in clinical medicine, are basically no more than a set of rules of inference. It is true that they do allow us to infer. But we infer, not from case to case, via a rule, but from the generalization regarded as a claim about some part of the world. The clinician is not just using the rCBF-dementia correlation to allow him to infer from one of the earlier patients to the 85th patient. Rather, he is arguing from his knowledge of the *entire* earlier group, that anyone who satisfies certain criteria of similarity to this group (in regard to rCBF, say) is also likely to display other similarities (dementia, say). The difference is more than merely one of emphasis. A rule is simply something that allows one to predict, to control. But a law-like assertion is something which demands to be *understood*; it characterizes a feature of the world for which a causal/theoretical understanding will be sought.

What bothers me most about the use of this analogy in this context, then, is its implicit restriction of the clinician to a mere use of rules in order to determine similarities of cases, thus apparently proscribing any reliance on his part on an understanding, however partial, of *why* such correlations as that between rCBF and dementia hold. Writing in more positivist days, Claude Bernard fell into a similar error. Despite our tendency to seek the "why" of things, experience teaches us (he asserts) that we cannot get beyond the "how":

> We know how water can be made, but why does the combination of one volume of oxygen with two volumes of hydrogen produce water? We have no idea. In medicine, it is equally absurd to concern oneself with the question "why". ([2], p. 80)

Bernard has proved to be quite wrong in this. First, we *do* know quite well why one volume of oxygen combines with two of hydrogen to form water; the structural theories of modern chemistry tell us why, in considerable detail. Of course, these theories are not the last word, but we have learned since Bernard's day not to worry about that, and to focus on the fact that they *are* a word, and a very powerful one in terms of disclosure of causal relations. More important, we have come to see just how important it is to concern ourselves with the "why" in medicine, first in order to understand what we do, and second, in order to do what we do even better. But more of that in a moment.

To summarize, the main reason why I oppose any appeal to legal analogies here is that they lead us to overlook that when a clinician uses an accepted correlation between a set of symptoms and a certain organic malfunction in order to diagnose, one can ask first: how well is the correlation established? and second: to what extent is it theoretically understood? Neither of these questions can be asked in the law-court, yet they bear vitally on the sort of inference the clinician is making, and in particular on the sort of assurance with which he may assert the correctness of his diagnosis.

3. THE MACHINE AS PARTNER

The second analogy I want to speak against is a far more familiar one, so familiar, in fact, that my efforts may well be in vain. Gedye takes over from cybernetics the description of a computer as "an artificial partner" for the clinician, one whose intelligent behavior, whose ability to develop new concepts (p. 107) will be such that the clinician will ultimately come to rely on his "partner's opinion" (p. 98).

I could start by saying that the computer is not an organism, that it does not exhibit intelligent behavior, that it does not have opinions, that it is a partner only in the sense in which a road-map or a working library is a partner of the traveller or the scholar. Metaphor is basic to human communication, and especially to the handling of new contexts. But the constant danger of metaphor is that it may be taken literally. The clinician who looks to his metallic companion for an opinion, and who really thinks it *is* an opinion, might tend to forget that what he is reading on the print-out is a transformation of a certain input designated to represent data (which *he* has determined to be data) according to rules of transformation which the programmer has chosen and built into the operation. The responsibility for the "opinion" in this case lies entirely on the diagnostician himself and on the programmer. There is no second autonomous intelligence entering in here, as a consultant might in a difficult clinical case. This is perhaps obvious, and I am sure it is obvious to Gedye, but one must be wary of the bewitchment of language here. It can too easily lead to a misdescription of the computer's real, and potentially very important, role.

In the case that Gedye describes, the computer plays, in fact, no role at all. He does not need, nor does he use, a computer to tell him how to organize the data from the first 84 patients in order to make a clinical judgment on the 85th. As the data accumulated, he could, of course, call on a computer, because of its first great advantage, its storage capacities. Or he *could* use file-cards; there is nothing that would demand the use of a computer. Nor is the manner of making the inference to what he calls "strong" assertions something that only a computer could help him to accomplish; pencil and paper will do. Of course, a computer might shorten the work.

But suppose we *do* decide to use a computer to facilitate the organization of the data, can we speak of a "simulation"? This is a complicated and controverted question. What would be simulating what? Presumably, the computer processes (I prefer the word 'process' to 'behavior' here) are supposed to simulate in some way the thought-processes of the clinician as he draws on his experience to make a judgment. What this amounts to is that the program is supposed to organize the data in order to allow an inference to be drawn somewhat as the clinician does. The program, let me insist again, need not be carried out by a computer; it can be done in long-hand. If the program *does* represent the steps followed by the clinician, then the long-hand computation or the computer process may be said in some sense to simulate what the clinician does.

However, I think that the simulation idea, whatever its validity, carries us in the wrong direction here. It depends for its validity on the assumption that the program really *does* represent the steps followed, implicitly or explicitly, by the practicing clinician. I think this would be quite difficult to show. Furthermore, it suggests that what we come up with will be the same as what the clinician would. But this is to underrate the value of statistical analysis, or if you like, the power of the computer. The *real* point of Gedye's proposal is not that it simulates what clinicians actually do, but rather that it suggests a program whereby clinical diagnosis could be made considerably more effective, quite independent of the question as to whether the program represents what clinicians are already doing and simply does it a little better. It would be better, then, to describe the role of the computer here as that of storage and organization. Properly programmed, it may enable the clinician to do something he was *not* doing before, namely to make numerical estimates based on explicit and exact correlations over an extensive material. It *extends* his abilities (as the thermometer or the X-ray machine does). But the fact that it does this successfully must not of itself lead us to say that it is *simulating* these abilities, no more than we would say that a thermometer "simulates" our sense of hot-and-cold.

What leads Gedye to the language of simulation is his rather special view of human knowledge itself. In a phrase reminiscent of Vico, he asserts that "we can only hope to understand what we ourselves have made, and as a consequence, that which we have *not* made only insofar as we can handle it 'as if' we had made it" (p. 97). The "technical psychology" of his title is an attempt to understand human behavior by means of analogies with the behavior of artifacts, such as computers, which we ourselves make. Their "psychology", so to speak, will enable us to understand ours.

This raises so many issues that it would need quite separate treatment. I will confine myself to two remarks. First, with all due respect to Vico and the other defenders of what is sometimes called the "*verum-factum*" thesis, I think it is false. It is not true that we understand all that we make; works of art are an obvious example. Nor is it true that we understand what is not of our making only by representing it as though it were. We understand molecules and chromosomes and volcanoes in a quite central sense of "understand" without in the least representing them as being of our making. It is true that we handle them in a language that is made by us. But if this were all the *verum-factum* thesis amounted to, it would be trivial, since all understanding requires language, and language is of human making. Though the concept *electron* is constructed by us, we do not understand electrons by somehow likening them to something we might have made. The materials of earth are too intractable and too unexpected to make this construal of the history of science and of the nature of scientific language at all plausible.

To say that we have "understood" how clinical judgment is made only when we have constructed a machine capable of simulating it is open to serious objection. We come to understand our own activities ordinarily by reflecting on them as they are exercised. We understand what it is to hate or to multiply by hating and multiplying and reflecting on what it is we do. There are certain activities, understanding itself preeminent among them, that are especially hard to grasp in this reflective way. Clinical inference is simpler; there is much that can be said about it (as this conference testifies) on the basis of direct and persistent scrutiny. One of the things that such scrutinizing permits is a formalization of the inferential processes involved. If this can be successfully done (as it can in the case of deductive inference), we can build the formalization into a computer program. But the artifact here is not the machine, it is the program. Or more basically, we do not understand the process here because we discover it in the machine; rather, it is because we understand it first that we *can* build it into the machine.

Let us suppose, however, that there is some activity of ours we cannot quite sort out; we do it successfully, but it is not clear, even after reflection, how we do it. So we gradually shape a computer program such that it can accomplish the same end (with the aid of an interpreter, of course!). This is, I think, what Gedye has in mind here. But how could one be sure in such a case that the computer program really *did* simulate what the clinician was doing?

Is it not possible that the two might be accomplishing the same, or similar, ends in quite different ways? How can one be sure that one is understanding clinical judgment as the *clinician* exercises it, in such a case? Only if one *already* knows that this *is* how he exercises it, I suspect, in which case the exercise in simulation will offer us no new light.

4. THE QUEST FOR CERTAINTY

Let us come now to the more distinctive feature of Gedye's paper, his quest for clinical certainty. Like Descartes faced by religious scepticism in the seventeenth century, who decided that nothing less than absolute certitude would suffice in the new system of philosophy he would propose, Gedye feels that the only way to "do the best we can" for the patient is to provide him with certainties, thus giving him a clear picture of the situation and "preparing the way for a rational acceptance of that situation and its implications" (p. 95). This is, in a way, an odd motivation here; one would have expected him to say that the main reason why certainty is needed in diagnosis is in order to enable a correct prognosis to be made and the proper therapy to be designed. What one tells the patient (though not unimportant) seems less important than getting him well again. Be this as it may, let us look briefly at the scheme he proposes.

Note first, however, that a high degree of certainty *is* frequently attainable in diagnosis. When the biopsy discloses certain kinds of abnormality or the throat-swab proves to carry streptococcus, the diagnosis can be definitive. The case Gedye considers is a special one, where a certain range of values of a measured variable, rCBF, show a correlation with a known family of disorders, dementia. It is assumed that a significant correlation has already been discovered by the usual means; one can proceed to refine this by checking the brain-regions separately and seeing which provide the best correlations with diagnosed certain cases of dementia. All of this is standard fare.

What Gedye proposes is to divide the range of the variable into three, on the basis of the data on the first 84 patients. In Figure 1, the variable is measured rCBF. The solid circles stand for patients diagnosed as affected by a form of dementia; the open circles represent patients diagnosed as "normal" (in the special sense of: not affected by dementia). The two boundaries in Gedye's ordering are X (marking the patient immediately below the lowest "normal" in the ordering), and Y (marking the patient immediately above the highest "abnormal" in the ordering). Thus every patient from X downwards on the scale has been diagnosed as "abnormal"; every patient from Y

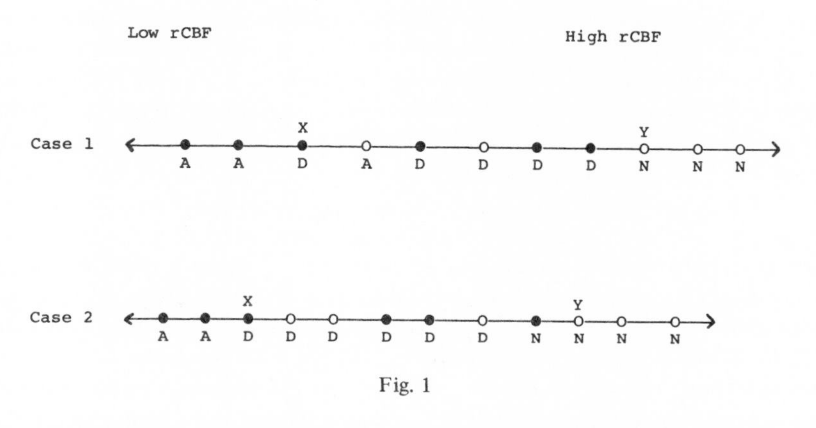

Fig. 1

upwards has been declared "normal". In the interval XY, some are affected by dementia; some are not. If the rCBF value of the 85th patient lies below X, Gedye claims that he can make a "strong" prediction, i.e., a prediction with certainty or near-certainty, that the patient will be found to be affected by dementia. And the degree of certainty can be ascertained, he suggests, by determining how well an abnormality in the 84th would have been predicted on the basis of the previous 83 cases, taking each of the 84 in turn to be the one to be predicted. After all, he argues, if we can determine retrospectively how many of the "strong" predictions made from the first 83 to the 84th would have been correct, this ought to give us an idea of how likely it is that a "strong" prediction in regard to the 85th will turn out to be correct.

How good a measure *does* this provide of degree of certainty? It follows from the way in which the measure is defined that all of the "strong" diagnoses of abnormality of the various patients taken as 84th, keeping the rest of the group the same, would have been correct, except for the "normal" immediately above X (case 1), *unless* where this "normal" is immediately followed by another "normal" (case 2). (The letters below the line indicate what the diagnosis would have been for this patient if he or she had been the 84th. A is abnormal, N is normal, D is doubtful.) If n is the number of patients in the sample with rCBF at or below X, the proportion of "strong" predictions of abnormality that would have proved correct in the 84th case (considering in turn each of the 84 whose records we already know as the 84th case) will be either $(n-1)/n$ (case 1), or n/n (case 2). There is no need to scrutinize each of the patient records in turn, by computer or otherwise. We know in advance exactly what the measure will be; it depends *only* on n, and on the diagnosis of the case which lies two above X. Both of these are known in advance.

This measure simply reflects the fact that we already *know* the 84-patient sample. We are certain to get the 84th right, in retrospect, unless it is at the boundary (which only one can be). All that this means is that the cases which *determine* our drawing of the X line will all, of course, lie below X if we carry out what Gedye calls a "judicial rehearsal" as to how they *would* have been diagnosed in advance on the basis of rCBF had they been, one at a time, taken as the 84th. But this gives us no idea as to how reliable the partitioning is. If only two cases of abnormality have as yet been found, the proposed measure in regard to our "strong" prediction for the next patient would be 2/2 in a case 2 situation, which makes no sense. One obviously cannot compare by this means the significance of the different brain-regions in regard to the correlations of their rCBF with dementia. The figures given by Gedye depend entirely on the contingent size of the sample one has taken for the particular brain region. One can increase his "degree of certainty" measure as much as one wishes by simply increasing sample size, thus increasing the number of "abnormals" below X. His "second-order rule" for combining the "best" brain-regions is, thus, useless.

What the statistician would want to know here is what the chances are of finding patients at different levels of rCBF below the present X-boundary who would, nonetheless, be diagnosed as normal by the clinician. There is no way to compute this from Gedye's schema. Nor can one compute from it the chances of finding a later rCBF value above Y which would, nonetheless, be associated with abnormality (clinically the more serious of the two possibilities). The X and Y boundaries will both tend to move backward as the sample-size is increased, so that the criteria for "strong" predictions will change, by how much we have no idea. Clearly, we have to fall back here on the normal statistical methods for estimating, first, the significance of the correlations, and second, the deviance curve. These methods will not give us sharp cuts between the certain diagnosis of abnormality, the doubtful case, and the certain diagnosis of normality that Gedye seeks. But there is, in fact, no method that *can* do that since such boundaries do not exist. There just isn't any sort of inference that could rule out the deviance curves we ordinarily expect in correlations of this sort. We can accept as a rule of thumb the claim that "a value of rCBF below any that has so far been discovered in a normal patient, where sample sizes are large, can safely be taken to indicate abnormality". As a working definition of 'safely' for the clinician, this will do. But we had better be quite clear that as a definition of 'certain' or even of 'almost certain', it will not.

There is a further complicating factor in making estimates of this sort, namely, the degree of theoretical understanding we have of the link between (in this case) the rCBF and the dementia. If we cannot see *any* causal link between them, the only grounds we have for the correlation are statistical. But if we *do* see the possibility of a causal correction, this obviously increases the degree of confidence with which we make our prediction. In particular, it tends to assure us that the statistical correlation is unlikely to be a chance one (something that on statistical grounds *alone* could never be ruled out). Since this is crucial for the understanding of clinical inference, and since it is commonly overlooked in standard treatments of Bayesian probability and of statistical inference generally, it seems worthwhile to end with a summary treatment of the distinction on which it rests.

5. AMPLIATIVE INFERENCE

Inferences can be of two kinds. The first, deductive, are such that the evidence is a complete and adequate warrant for the conclusion offered. If the premises are correct, we can be assured that the conclusion is correct. The second kind, which I will call by one of its many traditional names: 'ampliative' (rather than by Gedye's term, 'hypergnostic'), is where the conclusion goes *beyond* the evidence, and so can be asserted only in a provisional manner. A great deal of effort has gone in recent centuries to a more exact understanding of ampliative inference, of which clinical inference is one type.

It is crucially important to note that it may take two different forms. I will follow Peirce's way of drawing this distinction (see [7]). First, it may be *inductive*, that is, it may infer from a sample that certain correlations true of this sample will be true of a large class. Assumptions will have to be made about the reliability of the sample, but these assumptions can (up to a point) be tested. And other assumptions may have to be made about the uniformity of nature, or the like, though this will be a matter of debate. What is important is that the generalization arrived at introduces no new concepts. It goes, essentially, from "some A's are B's" to "all A's are B's", or from "a proportion P of the A's in this sample are B" to "P of all A's are B". Or in the form most familiar to the scientist, it will go from a finite number of points on a graph, each point representing a result, to a smooth curve, representing a generalization or a law. What permits the leap of inductive inference to be made is the assumption (perhaps incorrect) that our sample is a fair sample. It is *not* a matter of our understanding the causal processes involved, and seeing how these can be generalized.

This latter Peirce calls "retroductive" (or "abductive") inference. Here we move back from the evidence to a hypothesis which would in a strong sense *explain* the evidence. The hypothesis is not simply a law-like generalization from which the evidence could be deduced, as would be the case in inductive inference. Rather, it introduces novel explanatory concepts; it suggests theoretical structures, like molecules or chromosomes, which are in no way given in the original data; it suggests causal relationships, like those between planet and sun or proton and electron.

In the context of clinical practice, this distinction is altogether crucial. If an epidemiologist (to cite a recent paper – [6]) discovers a correlation between multiple sclerosis and distance from the equator, such that the risk of MS markedly increases with the latitude at which a person spends the first fifteen years of his or her life, an inference from latitude to proneness to MS will be an *inductive* one, as I have defined it. And one can use the apparatus of probability theory and sampling analysis to estimate the likelihood of such an inference.

On the other hand, suppose a correlation is noted between hypertension and heart disease. A clinician who uses hypertension as an indicator of heart disease is *not* just making an inductive inference from a large sample. He is also influenced by his understanding, imperfect though it may be, of the processes involved. In making his diagnosis he *is* then, (*pace* Claude Bernard), affected by the "why". He is much more confident about a correlation that he *understands* than about one which has just shown up in a computer cross-check. If the epidemiologist in the case just mentioned can show that there are intervening variables involving measles, and if the clinical researcher can show a probable causal connection between measles antibodies and MS, then the clinician can bring to his judgment about the likelihood of MS in a given case a far stronger form of inference than the original statistical correlation alone would permit.

The correlation between rCBF and dementia fits with what we know of the causal factors involved, although we are far from knowing just what the causal factors here are. Nonetheless, in making an inference from the correlation discovered in the first 84 patients to the next patient, the clinician is *not* just relying on the fact of the numerical correlation. If it had been between dementia and the length of surname, he would most likely have thought it a freak of the 84-patient sample that happened to be chosen. Thus, in his diagnosis, he is affected by the general state of understanding of the processes involved.

This is perhaps obvious, but it is absolutely vital to note it when questions come up about numerical estimates of likelihood or especially about computer simulation. For, as years of intensive work on ampliative inference have shown, retroductive inference does *not* lend itself (so far as can be told) to numerical probability estimates. We cannot attach significant numbers to the probability of our theory's (i.e., our understanding's) being correct. And thus we cannot attach an absolute probability number to an inference based upon this theory. Nor can we (so far, at least) construct a computer program which will simulate our understanding of causal processes and of our inferences from that understanding. A computer program for *inductive* or sampling inference is ordinarily easy to obtain, of course. But there is a real problem about retroductive inference. And since the clinician uses it all the time, this restricts the possibilities of simulation. This is *not* to say that the clinician would not learn much from computer-based correlations. But only to say that this is not *all* that goes into clinical judgments.

University of Notre Dame
Notre Dame, Indiana

BIBLIOGRAPHY

1. Bacon, F.: 1905, *Novum Organon*, P. Ellis and J. Spedding (transl.), G. Routledge and Sons, London.
2. Bernard, C.: 1927, *An Introduction to the Study of Experimental Medicine*, transl. by H. C. Greene, Schuman, New York.
3. Clifford, W. K.: 1872, 'On the Aims and Instruments of Scientific Thought,' in L. Stephen and F. Pollock (eds.), *Lectures and Essays*, Macmillan and Co., London.
4. Feinstein, A. R.: 1967, *Clinical Judgment*, Williams and Wilkins, Baltimore.
5. McMullin, E.: 1974, 'Logicality and Rationality: A Comment on Toulmin's Theory of Science,' *Boston Studies in the Philosophy of Science* **14**, 415–430.
6. Maugh, T. H.: 1977, 'Multiple Sclerosis: Genetic Links, Viruses Suspected,' *Science* **195**, 667–669.
7. Peirce, C. S.: 1933, 'On the Algebra of Logic', in C. Hartshorne and P. Weiss (eds.), *Collected Papers of Charles Sanders Peirce*, Harvard University Press, Cambridge, Vol. III (*Exact Logic*), pp. 104–157.

JOHN L. GEDYE

A REPLY TO ERNAN McMULLIN

My paper 'Simulating Clinical Judgment' was written in response to a request, from the organizers of this symposium, to give an account of my work in a way which would provoke philosophical comment. Having now studied Ernan McMullin's commentary on my paper I think I may claim to have achieved a certain measure of success in this endeavor, although McMullin's comments are, I must admit, not quite what I expected.

When in an *essay* one seeks, as I did, to cover an extensive territory in a relatively limited time, one gives, inevitably, a number of hostages to fortune. I am accordingly grateful to the editors of these proceedings for this opportunity to reply to the points McMullin has raised in his commentary.

McMullin titled his paper 'A clinician's quest for *certainty*' [my italics]; less elegant but more in line with my intention would have been 'A clinician's quest for a *reliable guide of action*'. I shall return to this distinction later.

The quotation from Simone de Beauvoir with which I began my paper was indeed intended to point up the clinician's dilemma when, aware of a world in 'shades of gray', he or she has, in the interests of the patient, to make a 'black and white' decision. McMullin, however, accuses me of 'seek[ing] to strip the clinician of his customary gray language and force him to "black and white" assertions (p. 115). Nowhere do I speak of '*forcing* "black" and "white" assertions' on the clinician. Clinical practice, by its very nature, *requires* "black and white" decisions – to invade or not to invade, for example – which must be made on the basis of a wide range of inputs of differing types, exhibiting a wide range of sensitivities. The problem is: 'how can the clinician arrive at a *just* decision in these circumstances, where the stark two-valued output of the judicial process seems so out of keeping with the rich texture of the input?'

McMullin, however, appears to have overlooked the solution I proposed to de Beauvoir's dilemma – that the 'gray' and the 'black and white' views need not be regarded as incompatible alternatives, that compatibility can be ensured if we can learn how to *construct* 'black and white' views appropriate to the occasion. Constructing such a 'black and white' view for some specific purpose does not mean that we have, in the process, to destroy our 'original copy' of the 'gray' view. One of the advantages of a computer-based approach

H. T. Engelhardt, Jr., S. F. Spicker and B. Towers (eds.), Clinical Judgment: A Critical Appraisal, 131–143. *All Rights Reserved.*

is that we can arrange to have this 'gray' view available to go back to whenever we wish.

Simone de Beauvoir's lamentation is surely the cry of one who – for lack of an alternative – is forced to use inappropriate existing conceptualisations, a familiar childhood experience. If there *is* an alternative to using arguments based on existing conceptualisations (and it was my purpose to suggest that there is), then it seems to me that, if it is at all possible for him to do so, the clinician has a responsibility to use arguments based on concepts constructed for the occasion, whether they be 'strong' or 'weak'. It is only when there is no alternative that he should accept having to use arguments based on existing conceptualisations.

At the present time, admittedly, the clinician has no practical alternative to working with existing conceptualisations, but this should not prevent us from exploring the implications of the fact that this need not necessarily be so in the future. If such explorations were to confirm the view for which I have argued – that this would be in the best interests of the patient – then, it seems to me, there is an obligation to move in this direction.

That McMullin finds the quotation from Clifford with which I end my paper 'sombre' is perhaps the most dramatic example of what I can only describe as the curious 'inversion' of my position which seems to run throughout his commentary. My reasons for ending with this quotation were twofold, and, I would have thought, obvious enough – first, its emphasis on 'scientific thought' as the *guide of action*, and second the recognition 'that the truth which it arrives at is *not* that which we can ideally contemplate *without error*,' that is: implies an unattainable certainty, but that which we may 'act upon *without fear*,' that is: in the knowledge that we are doing the best that we can do in the circumstances [my italics]. For Clifford, someone who saw this could not 'fail to see that scientific thought is not an accompaniment or condition of human progress but human progress itself'. As I might have anticipated, McMullin is unhappy about this outdated Victorian notion that 'scientific thought' is in some way to be equated with 'human progress'. This, I suspect, is a manifestation of what I might call the 'anti-evolutionary' tenor of his approach. My point will be made, however, if I am allowed to propose that we ought to consider whether, as Wittgenstein might have put it, "the legitimate heir to what used to be called 'scientific thought' can, in some sense, be equated with the legitimate heir to what used to be called 'human progress' ".

McMullin asks 'Are we to infer that the future of improvement of the clinical encounter. . . depends *entirely* on the extent to which the scientific rationality of that encounter can be *sharpened*?' [my italics]. McMullin clearly expects the answer 'no', but I am, at the risk of being misunderstood, strongly tempted to answer 'yes' to that question – if only to emphasize my belief that improvement of the clinical encounter depends on an increasingly precise demarcation of the boundary between its 'scientific' and 'non-scientific' components. I happen to believe that the non-scientific components are the more important, which is why I regard the task of developing scientific rationality to its limit to be so pressing – since it is only from the 'scientific' side that we can make 'progress' in demarcating the boundary between the 'scientific' and the 'non-scientific' components of the clinical encounter. My quarrrel with the avoidable use of probabilistic arguments is that it allows the clinician an invalid reason for avoiding important *non-scientific* issues.

In my paper I carefully avoided the phrase 'scientific method' because I do not believe that there is any such thing, I do not think that 'scientific thought' is distinguished by any particular 'method'. Yet McMullin attributes the phrase to me (page 115) and goes on to an account of Francis Bacon and his advocacy of a method that will necessarily lead to the truth, and which embodies a view of science as 'true causes known with certainty'. I agree that a reader might find *echoes* of this program in my paper, but my intention is so different that I am suprised to find McMullin proposing this parallel.

McMullin speaks of my 'uneasiness with the "almost certain", where the certain is what should be sought'. He goes on 'There is the same conviction that an orderly inductive method will do the trick where the naked force of the understanding (read: of the clinician's poorly organized intuitive skills) has failed'. My advocacy of "searching for the certain" is an expression of the view that we should not knowingly use a hypergnostic argument for which a counter-example exists (if we are willing to accept that this is what a clinician sometimes *unknowingly* does – because of his 'poorly organized intuitive skills' – then I accept McMullin's interpretation of my position).

In my paper I suggested ([2], p. 102) that when the word 'certain' is used in common discourse (as opposed to philosophical discussion, it seems), it seems to correspond to arguments for which there is no known counter-example – 'I was certain that she would visit me today'; 'Why? – because she has always visited me on Tuesdays in the past'.

Far from being uneasy about the 'almost' certain, I carefully pointed out that the use of the 'almost' was a way of marking the *possibility* of error in the absence of any precedent for error – 'she did not come because the public utilities were on a one-day strike this Tuesday – this had never happened to us before.' As Clifford put it 'the truth at which [scientific thought] arrives is *not* that which we can *ideally contemplate* without error, but that which we may *act upon* without fear' [my italics] – which is not to say that we will not sometimes make mistakes and get hurt, but that when we do we will be better able to accept our fate without recrimination, and to set out to see what we can do to avoid making the *same* mistake again in the future.

1. THE CLINICIAN AS EXPERIMENTER

I find relatively little to object to in McMullin's account of 'what the clinician's science can do for the clinician', but I am uneasy when he speaks of the clinician using a 'correlation' between rCBF and dementia, because it is the difficulties inherent in correlation – if by that one means the statistician's activity of computing correlation coefficients and the like – that led me to adopt the approach I describe in this paper, but more about that later.

2. THE LEGAL ANALOGY

I am, of course, only too aware that all analogies must break down at some point – unless, that is, we employ a definition of analogy which allows something to be analogous to itself. I take it, however, that in pursuing an analogy one is interested in just how far one can pursue it and in just why it breaks down. I am well aware of the many disanalogies between legal and clinical reasoning – which may be one of the reasons that the analogies do not appear to have been explored to any extent. However, I feel it to be very important for these analogies to be explored, provided of course, one is clear about what one is doing – it was for this reason that I carefully constructed examples of judicial processes – "judging 'black' or 'white' ", and "judging 'good' or 'bad' on the basis of 'black' or 'white' " to make the relevant meaning of 'judgment' and the scope of the analogy as explicit as possible. In spite of this McMullin accuses me of trying 'to explain the obscure by the more obscure' ([3], p. 119). I can understand him objecting to my examples and proposing alternatives, but I cannot see what he finds obscure about them – judging 'black' or 'white' seems to me to be about as simple an example of a judicial process as one can find – which is not, of course, to say that it is necessarily easy to understand.

There are several reasons why I feel it to be important to explore the analogies that *do* exist between clinical and legal thought, most of which are at least hinted at in my paper. Unfortunately I do not have space to pursue them all here, but one of the most important is that the operation of a legal system gives us an example of a decision-making mechanism created by human beings which has built-in capabilities for adaptation to changing circumstances. In my view this makes it particularly worthy of study in relation to clinical practice in a changing world. At the heart of the legal system's mechanism of adaptation is the principle of reasoning from case to case. It is my experience that clinicians do reason in this way – particularly when they are dealing with difficult cases – the very ones that give them the opportunity to demonstrate their highest professional skills. However, the process is certainly less explicit than in the legal counterpart – for the reasons, amongst others, of time scale that I discussed in my paper. I would even go so far as to say that a clinician's decision *can* sometimes constitute of itself a precedent which may serve as a warrant for later decisions – by the same clinician, or his colleagues. I have been witness to a situation in which a neurological surgeon argued that an invasive investigatory procedure should be undertaken on the grounds that he had decided on a previous occasion (some twenty-five years before) to undertake a comparable investigation in similar circumstances.

Much more important though, than such specific reasons for studying legal processes, is that I believe it to be particularly important to be aware of both the similarities and the differences between different types of human thought, and to seek to understand the reasons for them. In my paper I tried to suggest that many of the differences between clinical and legal thought might turn out, on closer examination, to be consequences of the different rates of turnover of cases and hence different information loads on the decision makers – so that the use of modern information handling technology in clinical practice might result in the clinician's office becoming progressively more analogous to a court of law.

McMullin goes on to challange what he calls 'the positivist supposition that scientific laws . . . are basically no more than a set of rules of inference'. It is not clear to me whether or not he is trying to imply that I hold such a view – all I can say is that nowhere do I talk about 'scientific laws' – I carefully avoided it. In my paper I went to some pains to distinguish (at least) two levels of activity – roughly corresponding to the 'Babylonian' and 'Greek' phases in the development of astronomy – an initial activity concerned with the detection of regularities, and a subsequent level concerned with the understanding of such regularities as had been detected at this initial level. From my experience in the field it seems to me that one of the most common obstacles to the development of a rational basis for clinical judgment is a *premature* attempt by clinical scientists to go to the second stage – to attempt understanding – without having first determined the form of what it is that has to be understood.

If, as McMullin claims, the clinician argues 'from his knowledge of the *entire* earlier group [of 84 patients]' presumably by means of the rCBF-dementia correlation, I would agree that he is 'using a generalisation regarded as a claim about some part of the world'. In my view this is just the trouble – the method of generalisation, whether it be the statistician's 'correlation' or whatever, itself embodies a structure (ultimately derived in this case, it would appear, from the mechanics of rotating bodies) which may not be – is indeed unlikely to be – relevant to the situation. Such claims are premature to say the least, it might be such a mismatch that puts us in a position of having to make probabilistic statements in circumstances where this could be avoided. Uncertainty could be the price we pay for 'putting the cart before the horse'.

The method of arguing from case to case, on the other hand, allows us to formulate a decision mechanism, useful in practice, the study of whose operation *may* allow us to go on to make generalisations about some part of the world, but this is a consequence of having found an inference procedure that works, not the reverse.

There is one further point that needs to be made here, and that concerns the role of the *entire* earlier group in the clinician's decision. If we follow the approach I advocated, this group of 84 patients sets, as it were, the bounds of our relevant universe at the time we prepare ourseves for the 85th. We draw on our knowledge of the entire group in order to determine appropriate conceptual thresholds to form concepts in terms of which we describe each of the 84 individuals, and in due course, the 85th. But having done this our argument is essentially from case to case – that is the function of the second order rule discussed in my paper. In my example 'similarity between cases' is determined by how many out of the four discriminating features chosen at rehearsal the case to be hypergnosed shares with each of the 84 preceding cases. Those cases with which the case to be hypergnosed shares the greatest number of features will be the most appropriate precedents for use in arguing from case to case.

McMullin then goes on to talk about the implicit restriction of the clinician to a 'mere use of rules' thus 'apparently proscribing any reliance on understanding'. I would have thought that it would have been clear from the whole context of my essay that understanding enters in – both in the choice of variables and in the attempt to account for whatever success is achieved by the judicial algorithm. The very existence of the investigatory technique of rCBF measurement and the fact that it was believed to be justified in these

84 patients indicates some grounds for believing in some kind of connection between rCBF and dementia. What is not known, though, is the *form* of that connection, and one cannot pre-suppose that it is such as to justify an 'a priori' belief that, say, the statistician's correlation coefficient or any other device for that matter will be appropriate to expressing it; although this might, in fact, turn out to be so. This is why our inferential procedure needs to be built on an 'external' as opposed to an 'internal' validation process, on what I called a 'judicial rehearsal'. It turns out in practice, in fact, that the form of the relationship between rCBF and dementia is more complicated than the simple example I used in my illustration would suggest. Abnormal patients may be characterised by both 'lo' and 'hi' blood flows, a disjunctive situation that my approach accommodates quite easily. It is not difficult to think up an explanation for this finding – in terms, say, of a redistribution of blood flow subsequent to some pathological event – so I would certainly not support the view that we cannot get beyond the 'why'. But at the present time, it seems to me, the temptation to do just this, without being clear about what it is that has to be explained, is a far greater danger.

3. THE MACHINE AS PARTNER

The first thing to be said is that I do not, of course, talk of 'the *machine*' as partner, but I will return to that point later. McMullin refers to me as 'tak[ing] over from cybernetics the description of *a computer* as "an artificial partner" for the clinician whose "intelligent behavior" will be such that the clinician will ultimately come to rely on his "partner's opinion" ' ([3], p. 121) [my italics].

The second thing I want to make clear is that as far as I am concerned I am not taking over anything from cybernetics – far from it. When I proposed the use of the term 'artificial organism' I was trying to find the best way I could, using natural language, to convey the nature of the activity in which I am engaged – if trying to be honest towards one's experience is being sentimental, then I accept the charge. So far I have found no better solution than to adopt the device of extending the language of interpersonal relations to include artifacts. In doing this I am doing something not very different from what the owner of a mass-produced automobile does when he ascribes a unique personality to his particular vehicle – I am reacting to the fact that all but the very simplest of artificial mechanisms demonstrate behavior that, however explicable in retrospect, in practice takes us by surprise when we meet it for the first time, but with which we gradually come to feel 'at home'.

If anything I am trying to emphasise how 'organism-like' machines are, not how 'machine-like' organisms are.

When I go on to refer to the artificial organism as a partner of the clinician I do so to indicate the nature of the relationship – the artificial organism is to be thought of not as a *substitute for* the clinician, but as a *partner of* the clinician. A partner whose opinion the clinician can, hopefully, come, as a result of experience, to value.

McMullin says that I use the term 'artificial partner' as a "*description of a computer*" [my italics]. I do not – I refer to the computer as a 'general purpose simulation device'. I do not call *it* an organism, but a means of simulating organisms – the computer allows us to create the partner, it is not itself the partner, any more than a child's constructional kit is the bridge he builds with it.

When McMullin refers to "intelligent behavior" he again gets things 'back to front'. What I was offering was a suggestion of what we might mean by 'intelligent' in this context – the behavior of an 'artificial organism' will be [clinically] intelligent insofar as a clinician comes to value its opinion on matters of clinical importance. I did not say 'rely on', I said 'value', but hopefully a valued opinion will come to be relied upon.

McMullin takes me to task for attributing an 'opinion' to the artificial organism, but this is a natural consequence of adopting personal language in the first place – the reasons for which I have already discussed. Providing a 'second opinion' is, perhaps, the characteristic function of a partner in clinical practice. We seek a second opinion when we are not sure about something, when we need input from an independent source. McMullin says that our source is not independent because we have designed it; it is, to use my phrase, an exteriorization of our own thought processes – a view with which I agree, of course. But does he really mean to suggest that we know ourselves so well that we have nothing more to learn? – he implies the opposite. The point is that while such an 'artificial partner' may not be better than we are when we are at our best, we are rarely at our best – we get tired, and we suffer from the other 'natural shocks that flesh is heir to'.

A student pilot quickly learns to distrust his orientational feelings and rely on his artificial horizon – if he doesn't, he may not live to tell the tale. The reason that pilots suffer from disorientation is that their occupation takes them into an environment with substantially more degrees of freedom of movement than that in which the human vestibular apparatus evolved. We can think of the artificial horizon as an artifact which allows the human visual system to 'take over' this orientational function from the vestibular system in the new environment. In the same way the 'artificial partner' can help the clinician to survive in an intellectual environment with more 'degrees of freedom' than that to which he has been accustomed – such as that created by the advent of CT scanning. In such a fluid world it surely cannot be in the best interests of the patient for the clinician to insist on 'going it alone' just to demonstrate his prowess. Would we fly with a pilot who adopted this attitude, however much we might applaud his bravado when he performs a 'look no aids!' stunt at an airshow?

McMullin says that in the simulation exercise I describe the computer plays no role at all. If by this he means that it would be possible to carry out the procedure I describe without a computer, then, of course I agree. In fact my own favorite 'primitive' state – to which I return in imagination when I have a difficult design problem to solve – is a stretch of sand, a pile of stones, and a stick, and, of course, *unlimited* time. Ay, there's the rub – the reason we use the computer is to correct the temporal mismatch between our 'primitive' situation and the world in which we live. One of the basic themes running through my paper is that we should be asking to what extent seemingly disparate human judicial activities are, at a fundamental level, more alike than we think – the differences being attributable to differences in the time available for a solution to be arrived at in the real world.

We now come to the question of simulation, which McMullin regards as a "complicated and controverted" question (and having read his commentary I agree that it can certainly be made to appear so) – he asks "what would be simulating what?". The straight – and somewhat oversimplified – answer to this question is that the clinician is simulating himself, or more explicitly the clinician at one point in time simulates himself at an earlier point in time (the last time he faced a similar problem) – since the most appropriate precedent for a particular clinical situation is the most similar previous situation that he can find. The result of this process is the simulation. Note though that the simulation goes beyond the real life situation insofar as it embodies what I referred to as a judicial rehearsal, a device which allows each previous patient to be judged in the context of the rest, even though this was not possible in real life.

Perhaps part of McMullin's difficulty is attributable to a confusion between the verb to simulate meaning 'to *create* a simulation of' as opposed to 'to *be* a simulation of'. I think it is clear that in my paper 'simulating clinical judgment' refers to the activity of *creating* an artificial organism, such that the clinician will come to value its opinion on matters of clinical importance'.

McMullin feels that to speak of simulation carries us in the wrong direction because 'It depends for its validity on the assumption that the simulation really *does* represent the steps followed, implicitly or explicitly, by the practicing clinician' ([3], p. 122). McMullin seems to have missed the significance of the point, that I went to some lengths to make, that simulation is a means of innovation, the *results* of which may lead to understanding. We observe some behavior and we set out, using well-defined constructional techniques to make something that will exhibit similar behavior, having decided in advance how we will judge similarity or difference. In the case I discuss in the paper the aim is to end up with a simulation whose *performance* can be directly compared with that of a clinician. In this case similarity or difference is judged by ability to accept a similar range of inputs and produce a similar range of similarly acceptable outputs – the judicial process referred to earlier.

Why simulate the clinician if we want to be able to do better than the clinician? When we simulate the clinician we simulate those aspects of his behavior that we would like to encourage – the clinician 'at his best'. We are interested in what constructional techniques are *sufficient* to achieve this result. This knowledge may enable us to see how we can do better than the clinician can do – for the same kind of reason that, within its domain, a special purpose computer can often do a better job than a general purpose computer – we pay a price for generality. It is worth noting here that the interest of the ancient geometrical problem of trisecting an angle is not so much to see if it can be done, but to see if it can be done *using only a straight edge and compasses.*

McMullin is certainly on the right lines when he links my position with that of Vico – I have felt for some years that I would like to know a lot more about Vico than I do – but he is not on the right lines if he thinks that I am advocating a "verum-factum" thesis which necessarily applies to *all* human knowledge. I take the view I do because I have found it a useful basis for my day-to-day activities in clinical research, and I readily admit to being unable, at present, to conceive of a situation in which I could honestly say that I understood something that I could not handle in this way. This is, of course, to define what I mean by understanding in the course of *my* work, but I am not at all sure that if I were a vulcanologist I would feel able to claim to understand a volcano beyond the extent to which I could handle it in like manner.

4. THE QUEST FOR CERTAINTY

I have already drawn attention to the fact that it would be more appropriate to refer to my interest as a quest for a 'reliable guide of action'. But, be that as it may, McMullin quotes me as proposing that "the only way to 'do the best we can' for the patient is to provide him with certainties". I was proposing nothing of the kind, in the relevant passage ([2], p. 109). I was emphasizing the importance of the distinction between 'strong' and 'weak' arguments. I indicated that if we accepted this we could claim (almost) certain knowledge about *some* things, but that it required us to refrain from such claims with respect to other things. The approach allows us to partition our knowledge into what we do and what we do not claim to know something about. I was proposing that it was by accepting this position, the *partitioning* of our knowledge, that we would be led to do the best we could for the individual patient. We offer the patient not certainty in McMullin's sense but guidance based on an honest appraisal of the evidence. To treat all arguments as 'weak' where there is a possibility of our being able to put forward a 'strong' argument is, in my view, dishonest.

In discussing my simulation exercise McMullin says 'It is assumed that a significant correlation has already been discovered by the usual means' ([3], p. 124). My approach involves no such assumption, it was to avoid having to make such assumptions – for the reasons discussed earlier – that I developed the approach in the first place. It is clear from his account of the basis of my approach that McMullin has completely misunderstood the grounds on which an (almost) certain claim is based. These grounds are nothing to do with the thresholds as such, rather they derive from the fact that the mechanism I describe partitions the set of possible claims into those for which there is no known counter-example and those for which there is at least one known counter-example.

The fact that a single brain region will have been correct in only $(n\text{-}1)/n$ of the 'strong' predictions for the previous patient is, of course, the very reason that we need to recruit more than one brain region on which to base our judgment, and why our second order rule involves an expression of the form of 'possessing x, or more, out of y features as determined by first order rules (where $x < y$)'. This can be thought of as a method for creating a class based on a precise definition of what Wittgenstein called a 'family resemblance', and which Atkin has recently formalized in his 'q-connectivity' ([1], pp. 1–209).

It is necessary, of course, to settle on values of x and y, and we do this at our judicial rehearsal – which allows us to choose these values in such a way as to maximize the number of successful strong claims that are put forward at the step prior to hypergnosing the new patient.

There is, of course, no way in which we can assign a probability (in the technical sense) to being wrong when we apply this rule to the next patient.

I deliberately raised this issue in the form of a question as to how we might defend ourselves against a charge of irresponsibility, rather than as a question about probabilities relating to a future event. What we *can* do is see how often we would have been wrong if we had operated on this basis in the past. If such an experiment gives a non-zero error rate we can re-apply our fundamental principles, as appropriate, and continue until we have no known counter examples. It is then, and only then, that we claim (almost) certainty for our 'strong' arguments. We are, in other words, adopting a 'null' method – we do whatever is necessary to eliminate counter-examples. If this means that we end up able to say very little, then so be it.

As far as 'retroductive inference' is concerned, I have already pointed out that this operates when the clinician decides to subject the patient to the investigation in the first place. He would not, without reason, be likely to propose that such a feature as the length of the patient's last name be included in the decision data base less he had good reason, but if it 'survived rehearsal' he might come to depend on it like anything else. But it would, of course, raise the question 'why?' – and pursuit of this might, for example, lead to some connection with ethnic background, as reflected in length of lastnames.

At the end of my presentation Dr Cassell made a comment which helps to put my whole thesis in a nutshell, he said "what you have done is to simulate the *bad* clinician". He was referring to the fact that if one froze the 'artificial partner' at, say, the 84 patient level and allowed it to operate without further revision, it might perform fairly well for a while if the situation was fairly stable, but it would begin to deteriorate as soon as its experience (the 84 patients) became outdated. It would be behaving like a clinician who thinks he knows all there is to know about a particular condition and so blinds himself to the contrary evidence.

It is, of course, inherent in the approach I have described that the system will adjust itself after each new patient is added and what the decision should have been is known. As the system acquires counter-examples it will, as it were, self-destruct – the number of 'strong' arguments will go down and the number of 'weak' arguments will go up. When this happens it will be a sign that it is time to look for a better source of data for whatever decision it is that we are trying to make. We thus see that the approach is essentially an evolutionary one, and hence appropriate to clinical practice in a changing world.

Baylor College of Medicine
Texas Medical Center
Houston, Texas

BIBLIOGRAPHY

1. Atkin, R. H.: 1974, *Mathematical Structure in Human Affairs*, Crane and Russak, New York.
2. Gedye, J. L.: 1979, 'Simulating Clinical Judgment (1): An Essay in Technological Psychology', this volume, pp. 93–113.
3. McMullin, E.: 1979, 'A Clinician's Quest for Certainty', this volume, pp. 115–129.

PATRICK SUPPES

THE LOGIC OF CLINICAL JUDGMENT: BAYESIAN AND OTHER APPROACHES

Not many years ago it would have seemed impractical, if not impossible, to have physicians and philosophers engaged in dialogue about the logic and nature of clinical judgment. The philosophers would have been unwilling or unprepared to think about matters that on the surface seemed far removed from classical philosophical problems. Physicians on their part would have been wary of entering into the labyrinth of methodological issues dealing with the relation between judgment and evidence. Now it seems wholly natural to have such an interaction and to have a conference that focuses on clinical judgment, with physicians and philosophers doing their best to interact and to understand each other's problems and methods.

I am sure that a difficulty for all of us is not to get carried away with expounding the technical subjects on which we are now working and to strive to communicate at the appropriate level of generality and simplicity. I know from experience that medical talk about any specialized area of disease can almost immediately get beyond my competence and knowledge if the full clinical details are presented. Over the past several years I have had the pleasure of talking about matters that are generally relevant to this conference with my colleagues in the Stanford Medical School. A number of these conversations have been with members of the Division of Clinical Pharmacology. I have been pleasantly surprised at my ability to get a sense of the problems they consider important to attack, even though the detailed terminology and data of clinical pharmacology lie outside areas of knowledge about which I claim to have accurate ideas. I think that the same goes for my own areas of special knowledge. It would be easy enough for me to raise particular questions in the foundations of probability or decision theory that are of current concern to me and that have some general relevance to the theory of clinical judgment, but that would be too specialized and esoteric for detailed discussion in this context. No doubt I shall not be able to be totally austere in this forbearance and will occasionally at least allude to current technical interests of my own that have potential relevance to the topic of this conference.

There is of course another danger – a practice of which philosophers are often guilty – that what I have to say could be formulated in such a general

H. T. Engelhardt, Jr., S. F. Spicker and B. Towers (eds.), Clinical Judgment: A Critical Appraisal, 145–159.

way that it would not really be of interest to anyone, perhaps because the ideas in their most general form are already widely familiar.

With these considerations in mind I have divided my paper into four sections. The first deals with probability and the general foundations of statistical inference, with attention focused on the Bayesian approach. The second section enlarges the framework of probability to that of decision theory by introducing the concept of the value or utility of consequences. Unlike many applications of modern decision theory to scientific research, the application to clinical judgment seems especially natural and appropriate. The third section deals with models. The main point here is that a general theory of decision making is no substitute for particular scientific understanding. The fourth and final section deals with what seem to be some of the perplexing problems of data analysis in medicine, at least from the perspective of an outsider who has had more problems and experience with data analysis in other areas than he cares to think about.

I. PROBABILITY

Among fundamental scientific concepts, that of quantitative probability is a late arrival on the scene, as my colleague Ian Hacking has shown in a splendid monograph on the emergence of the concept in the 17th century [4]. The theory of statistical inference is even more recent, and is really only a product of the 20th century. Given the long and developed history of medicine, reaching back for thousands of years, it is not surprising that the recent concepts of probability and statistics have as yet had little impact on the practice of medicine.

Moreover, some of the foundational views of probability do not in any natural way lend themselves to the clinical practice of medicine. One important and fundamental approach to probability has been to emphasize that probability always rests on the estimation of relative frequency of favorable cases to possible cases in some repeatable phenomena, of which games of chance provide paradigm examples. The long history of emphasis in medicine on the diagnosis of the individual case does not easily lend itself to this relative frequency view of probability.

Fortunately, there is an equally persuasive and important view of probability as expressing primarily degree of belief or, as it is sometimes put for pedantic purposes, degree of partial belief. In ordinary talk, most of us consider it sensible to ask what is the probability of rain tomorrow. We have even come to expect the evening TV news to provide a numerical estimate. When the forecaster says that the chance of rain tomorrow is 60 percent, he is not using in any direct way the relative frequency approach but is expressing his degree of belief even if he does not himself explicitly use such language.

Physicians, it seems to me, generally do not need any persuasion about the importance and value of the expression of a degree of belief as an approach to probability. This approach is often called Bayesian because of its early lucid formulation as a foundational viewpoint by the Reverend Thomas Bayes in the 18th century [1].

The centerpiece of this approach is Bayes' theorem, which says that the posterior probability of a hypothesis, given evidence, is proportional to the likelihood of the evidence, given the hypothesis, times the prior probability of the hypothesis itself. If consequences are ignored, then the maximum of rationality that follows from Bayes' theorem is that we should act on the hypothesis that has the highest posterior probability.

Let us examine some of the difficulties in a direct application of Bayes' theorem to clinical practice. A general problem is the unwillingness of many physicians, on the basis of temperament and training, to put themselves in an intellectual framework that calls for probability judgments in diagnosing a patient's illness. It seems to me that there are two good intellectual reasons for this resistance on the part of physicians. The first is skepticism that a mechanical or semimechanical algorithm can be as effective in assessing a patient's state as the intuitive judgment of an experienced diagnostician. It has not been part of the long tradition of clinical practice to attempt numerical assessments of any kind, really, of a patient's state, and an experienced clinician can be skeptical that a reduction to a numerical statement is feasible. There are many kinds of decisions or judgments we make that are not easily or naturally reduced to verbal rules, let alone quantitative rules. Perhaps one of the simplest examples that has received a good deal of study in experimental psychology is the way in which we recognize faces or familiar smells. Either in the visual case of face recognition or in the olfactory recognition of familiar smells the verbal descriptions we can give of the evidence on which our decisions of recognition are based are extremely poor and vague in character. A reduction to explicit rule of recognition procedures in either of these relatively simple but familiar domains would probably be unworkable for even the most articulate. Similarly, so the argument goes, the intuitive judgment of clinicians is based upon a depth and range of experience that cannot be reduced to explicit rules.

The second major argument against Bayes' theorem is that even in arenas where explicit data, for example, of a laboratory sort, are being considered and a framework of explicit concepts is being used, there is often no natural and nonarbitrary way to incorporate new objective evidence within a feasible application of Bayes' theorem. In this case the evidence is explicit and the data are objective, but we do not have explicit rules for calculating likelihoods. We especially do not have such rules when we suspect there is strong probabilistic dependence among various parts of the evidence being considered.

I respect both of these arguments in an essential way. As far as I can see, there is no reason to believe that the time will ever come when we can have any simple direct mechanical application of Bayes' theorem or similar statistical tools to provide a satisfactory but automatic diagnosis of an individual patient's illness. This does not mean that I am against pushing the use of Bayes' theorem and other similar methods as far as we can and indeed insisting on as many studies as possible of their feasible application. A number of studies, in fact, of the use of Bayes' theorem in medical diagnosis have already been made, and several with quite positive results. For example, Warner et al. [13] incorporated Bayes' theorem in a computer program that was used in the diagnosis of congenital heart disease. The program that applied Bayes' theorem classified new patients with an accuracy close to that of experienced cardiologists. I shall not attempt to survey here the number of other excellent studies in this direction. A good brief survey is to be found in Shortliffe [9], who is sympathetic to what has been demonstrated thus far but who is at the same time dissatisfied with a Bayesian statistical approach as being anything like the final word.

The remaining three sections of this paper deal with broad concepts that I think are necessary to augment a Bayesian approach to clinical judgment. Before turning to these matters, I do want to emphasize that I have made no attempt here to enter into the deeper technical developments of the Bayesian theory of statistical inference or alternatives that have been extensively studied by mathematical statisticians over the past several decades.

II. EVALUATION OF CONSEQUENCES

The standard Bayesian application often emphasizes only the assessment of beliefs and how these beliefs change with the accumulation of new evidence. The general theoretical setting, however, of decision theory has emphasized the fundamental place not only of belief but also of evaluation of consequences. It is my impression that the current literature on clinical judgment in medicine has placed much more emphasis on methods for assessing beliefs consistently and rationally than it has on assessing rational methods of evaluating the consequences of the decisions taken. There is undoubtedly a sound intuition back of this emphasis. If all the evidence is in, it often seems clear enough what action should be taken and what the anticipated consequences of the action will be. For example, if a patient is diagnosed with extremely high probability to have a particular infectious disease for which there is a standard, highly specific treatment, and, moreover, the probability is low that any known side effects of the treatment will have deleterious effects on the patient, then the main task of decision making is over. The consequences of deciding to prescribe the standard treatment seem obvious and do not require extended analysis.

The difficulty, of course, is that this kind of clear situation seldom obtains. Moreover, in the context of modern medicine, a new factor has arisen which has already led to some emphasis being given to problems of evaluation of consequences. This is the problem of holding down the cost of laboratory or physical tests [3].

In practice, of course, physicians do automatically attach some evaluation of consequences, including the cost of laboratory tests, for if they did not, the rational decision would always be to require as many laboratory tests as possible in order to maximize the evidence available in making a diagnosis. Reasonable rules of thumb no doubt were appropriate and proper in the past. With the much more elaborate possibilities available today, and with the costs rapidly rising of the more sophisticated laboratory tests, it also seems appropriate that more elaborate tools of decision making will come to have a natural place in clinical judgment.

It is safe to predict that, with the national concern to control the costs of health services, attention to the problem of costs just mentioned will be a major focus of both theoretical and practical work on clinical judgment. I would like, however, to focus on some different issues that arise from the evaluation of consequences and that have possible implications for changing the traditional relation between physician and patient. What I have to say about these matters is certainly tentative and sketchy in character, but the issues raised are important and, moreover, the tools for dealing with them in a rather specific way are available.

The issue I have in mind is that of making explicit the possible consequences of a decision taken about medical treatment on the basis of clinical judgment. Traditionally, no very explicit model of evaluation is used by the physician either for his own decision making or for his consultation with the patient about what decision should be made. It is a proper part of the traditional relation between physician and patient that with certain unusual exceptions, the final decision about treatment is the patient's and not the physician's. On the other hand, it is a part not only of traditional but also of modern medicine for the vast majority of patients to accept the treatment that is preferred by the physician. I have not seen any real data on this question but it would be my conjecture that in most cases the physician makes relatively obvious to the patient what he thinks is the preferred treatment. In cases of certain risky operations or experimental drug treatments, etc., almost certainly there is a much stronger tendency to lay out the options for the patient and to make explicit to him the risks he is taking in the decisions he makes.

It is especially the decisions that have possibly grave negative consequences to the patient that suggest a more explicit analysis of the decision process. To provide a concrete example for discussion, let us consider the following highly simplified case. I hope that you will bear with the obviously oversimplified character of my description. Let us suppose that the patient is one with a serious heart condition. He is presented by the heart surgeon to whom he has been referred with any one of three options: bypass surgery, continual treatment with drugs and without surgery, no treatment of any sort. To carry through an explicit decision model of a quantitative sort, the patient needs now to be presented data on possible consequences of each of the three actions, together with the medical evidence on his heart condition. According to the standard expected utility scheme, the patient should then select the medical treatment that will maximize his own expected value or expected consequences or, in other terminology, expected utility.

Satisfaction of the conditions required to make an expected-value model work well do not seem easy to come by even in the most clearcut clinical situations, and consequently I would like to examine in some more detail the feasibility of using such a model at all.

The problems that arise naturally separate into two parts. One part concerns the ability of physicians to make quantitative assessments of possible consequences of treatment as well as of the current true state of the patient as inferred from the evidence available. We can question whether the state of the art and of the science can bear the load of such quantitative assessments at the present time. We may certainly want to hold the view that assessment of individual cases in such a quantitative fashion is not practical. I will return to this point in a moment. The second part concerns the ability of the patients to absorb the data concerning the options presented to them by the physician. The number of patients who feel at home with probability calculations or the concept of expected value is very small indeed. Anything like a routine quantitative application of the model seems totally impractical for all but an extremely small segment of patients, at least at the present time. It may properly be claimed that the detailed quantitative assessment of evidence or of consequences is as complicated a technical topic as the laboratory tests called for by the clinician, and the ordinary patient is simply not competent to deal with a quantitative decision model even when its application is of great personal consequence to himself.

There is a second approach that seems a good deal more promising in the present context of medical practice and the expected knowledge of patients. This is to move the development and analysis of a quantitative decision-making model from the level of the individual case to statistical analysis of a large number of cases. It is certainly true, for example, that the consequences will vary enormously from one patient to another, not only because of his physical condition but also because of his age, his wealth, his family responsibilities, etc. On the other hand, there are clearly four consequences that dominate the analysis of the full nexus of consequences, namely, (i) the probability of recovery, (ii) the probability of death, (iii) the probability of serious side effects in terms of medical consequences, and (iv) the expected cost in terms of types of treatment. In summary, the direct medical consequences and the direct financial costs of a given method of treatment are the most important consequences, and these can be evaluated by summing across patients and ignoring more detailed individual features. This does not mean, for example, that in assessing the consequences of treatment we ignore the age of the patient, because this is part of the evidence and should go into the assessment of the consequences for the given state of the patient.

The unconditional or mean assessment of the consequences of particular treatments is a relatively straightforward piece of data analysis. The situation is quite different, however, if we want to make the appropriate conditional assessment – conditional upon the variation in relevant parameters of the patient's state at the time of treatment. The complexity of estimating the joint probability distribution of various consequences – or symptoms – is admitted by almost everyone who has considered the problem.

What seems desirable at the present time is development of an actuarial approach to both the state of health and the consequences of treatment. Such an actuarial analysis could serve only as a guideline in the treatment of individual cases, but useful information about medical decisions by individual physicians or groups of physicians could be obtained.

Let me give an example. A number of studies [6, 8, 7, 5, 10, 2] have shown that there is a definite tendency for nonspecialists to prescribe more antibiotics than are required by patients' conditions. Studies of this kind provide an excellent way of cautioning physicians to consider carefully the clinical basis of any prescription of antibiotics but do not attempt at all to provide an algorithmic or mechanical approach to clinical diagnosis.

III. MODELS

Although the merits of Bayesian and related methods of inference can be defended as practical tools that can be brought to bear on real problems of clinical judgment, it is important to emphasize that such methods are no panacea and do not provide in themselves a scientific foundation for medicine that is in any sense self-sufficient. It is quite true that there are areas of medicine that are clinically important and that do not at present have a thoroughly developed theory. My better informed friends tell me that this is true of more areas of clinical medicine than I would be naturally inclined to believe, but I certainly won't venture to give details on this point and simply take it as an assumption that it is easy to draw distinctions between various areas of clinical medicine. The distinction is concerned with those that have a practically applicable theory and those that do not. Those that do not, it seems to me, can especially benefit from the application of Bayesian methods, but this benefit should not obscure the need to continue the development of a more adequate scientific foundation. Moreover, the development of that better scientific foundation surely does not depend on any direct application of statistical methods, but rather on the creative development of new scientific concepts and theories. Explicit scientific models of the relevant biological phenomena must remain a goal, I would suppose, in every area of clinical medicine.

As such models develop, we should be able to fold them into a general framework of statistical inference and there should be no natural conflict between the use of newly developed models and the stable data-based inferences of the past. There will, of course, be the practical problem of assaying the relative weight to be given to past experience, on the one hand, and, on the other, the relative weight to be given to the new scientific models of the phenomena at hand.

There is nothing about this situation that is distinctive and special to the problems of clinical medicine. A similar tension between past experience and the urge to develop deeper scientific models is to be found in every area of applied science. Salient examples that confront us every day and that are nearly as important as the problems of medicine are to be found in economics and meteorology. Moreover, we are all conscious of the difficulties of developing adequate scientific models either of the economy or of the weather, but the thrust to do so is deep and sustained, just as it is in modern medicine.

The introduction of more general and more powerful scientific models in various clinical areas seems to me to generate an interesting and important problem of differentiating the future possibilities. On the one hand, the scientific thrust is to make the clinical diagnosis of a patient ever more algorithmic. Although, as I have argued above, we shall never pass beyond the need for clinical judgment, it is still important to recognize that what we can see in the reasonably near future for different parts of medicine presents quite a different picture. In the diagnosis of infectious diseases, for example, we might expect to get nearly algorithmic laboratory and computer-based procedures – at least I will venture that conjecture. On the other hand, in the diagnosis and treatment of psychiatric disorders we may anticipate in no reasonable future a scientific model of sufficient depth and generality to provide anything like algorithmic diagnostic procedures. I wish that I were competent to give a survey of the various areas of medicine and to conjecture what we might expect along algorithmic lines. I would be enormously interested in hearing more informed opinion than my own about this matter.

It is also not clear what we may expect from models that derive from work in artificial intelligence. The MYCIN program of Shortliffe [9], for example, has many attractive features and is potentially a diagnostic aid of great power, but it is still rather far from being ready for practical daily use, even in the sophisticated setting of a teaching hospital.

IV. DATA ANALYSIS

Let me begin my remarks about data analysis with a tale of my own. A couple of years ago we embarked on collecting a large corpus of spoken speech of a young child. The mother of the child was hired as a half-time research assistant to spend twenty hours each week creating a properly edited computer file on which she transcribed an hour of the child's spoken speech for that week. In something over a year of effort on the part of the mother we obtained a corpus of more than a hundred thousand words of the young child from the age of approximately two years to three and a half years. We then engaged in elaborate computations regarding structural features of the speech, especially an elaborate test of a number of different generative grammars and model-theoretic semantics and of developmental models of grammatical usage. We estimated that at the conclusion of our elaborate computational analysis of a number of different grammatical models we had probably done more explicit computing than had been done by all the linguists of the 19th century working on all the languages examined. Even so, the piece of data analysis we did on a child's speech seems trivial compared to the overwhelming problems of data analysis in clinical medicine. In just one large clinic, consisting of, say, a hundred doctors and their support staff, the data flow is like a torrent and the problem of providing sensible analysis appears almost overwhelming.

However, it seems to me that there is much that is constructive that can be done and that can provide important supporting analyses for those responsible for final clinical judgments.

The first fallacy to avoid in attempting this is the philosopher's fallacy of certainty. There is no hope of getting matters exactly right and organizing a body of data that will lead to certain and completely reliable conclusions about any given patient. It is important to recognize from the start that the analysis must be schematic, approximate, and in many cases crude.

Second, the penchant of many social scientists and applied statisticians for experimental designs must be recognized in this context as a romantic longing for a paradise that can never be gained. Just because it is not practical to impose experimental methods of design or parameter variation on the flow of patients through a standard medical clinic, it does not follow that any quantitative approach to causal analysis must be abandoned. Interestingly enough, some of the very best modern methodology has been developed and is being used by econometricians dealing with data that are totally inaccessible to experimental manipulation. Moreover, the statistical analysis of data was first used in a massive way in the least experimental of the physical sciences, namely, astronomy. We should no more despair of the severe limitation on experimentation in clinical medicine than astronomers of the 18th century despaired at the absence of the possibility of astronomical experiments. I want to make this point explicit because it is one that I have spent a good deal of time on in casual argument with philosophers and statisticians whose opinions I respect but do not agree with. One of the most sophisticated and significant applications of probabilistic and statistical analysis to the identification of causes was Laplace's method of what he termed "constant" causes. He used these methods to attack the subtle problems involved in the effect of the motion of the moon on the motion of the earth, to analyze the irregularities in the motion of Jupiter and Saturn, and to identify and consequently to explain the mean movements of the first three satellites of Jupiter. These are classical results from the late 18th century, but one has to look far and wide in the entire history of science to find *experimental* results of comparable conceptual and quantitative sophistication.

The third point amplifies some earlier remarks about joint probabilities. Even crude approximations to the joint distribution of causes or symptoms would, I would conjecture, be a definite methodological step forward in the analysis of clinical data. In this case I am returning to my earlier remarks about looking at large numbers of cases and applying the results as guidelines for considering individual patients. It is a first lesson in elementary probability theory that from the (marginal) distributions of single properties it is not possible to infer the joint distribution of the properties. The estimation of these joint distributions is a complex and subtle affair, but it is my belief that in many cases even relatively crude results would lead to clinical insights of considerable interest. To show that the theory of approximate probabilities can itself be given an elementary treatment in qualitative terms, I have added to this paper an appendix setting forth in outline the axiomatic foundations, but I do not want to disturb the main line of my general discussion by considering the technical details here.

The fourth point concerns the great importance of considering alternative hypotheses or causes to provide a perspective on the identification of the most likely cause. Consideration of alternative hypotheses is natural within a Bayesian or a classical objective framework of statistical inference and is a matter that is old hat to statisticians, but it is not only in medicine but in other parts of science as well that the explicit formulation and analysis of the data from the standpoint of alternative hypotheses are far too often not undertaken. I do not mean to suggest that good clinicians do not range over a natural set of possibilities in diagnosing the illness of a patient, but rather that, in the kind of quantitative data analysis based on many cases that I am advocating as a general intellectual support, analysis is often not adequately presented of the support the data give to alternative hypotheses or causes.

To show that the philosophical thrust of this last remark is not new, as indeed is true of most of the other things I have had to say, let me close with a quotation from Epicurus's letter to Pythocles, written about 300 B.C. soon after the very beginning of philosophy as we know it. Epicurus's remarks are aimed at our knowledge of the heavens or, more generally, of the universe around us but they apply as well to the focus of our present discussion.

For this is not so with the things above us: they admit of more than one cause of coming into being and more than one account of their nature which harmonizes with our sensations. For we must not conduct scientific investigation by means of empty assumptions and arbitrary principles, but follow the lead of phenomena: for our life has not now any place for irrational belief and groundless imaginings, but we must live free from trouble. Now all goes on without disturbance as far as regards each of those things which may be explained in several ways so as to harmonize with what we perceive, when one admits, as we are bound to do, probable theories about them. But when one accepts one theory and rejects another, which harmonizes just as well with the phenomenon, it is obvious that he altogether leaves the path of scientific inquiry and has recourse to myth. ([11], p. 19.)

Stanford University
Stanford, California

APPENDIX
APPROXIMATE PROBABILITIES

The first step in escaping some of the misplaced precision of standard statistics is to replace the concept of probability by that of upper and lower probability. The concept of upper and lower probability provides an appropriate way of handling approximation. What I have to say is developed in more detail in Suppes [12], and I shall only sketch the results here.

To begin with, let X be the sample space, $\mathscr{F}$ an algebra of events on X, and A and B events, i.e., elements of $\mathscr{F}$. The three essential properties we expect upper and lower measures on the algebra $\mathscr{F}$ to satisfy are the following:

(I) $P_*(A) \geqslant 0$.

(II) $P_*(X) = P^*(X) = 1$.

(III) If $A \cap B = \emptyset$ then

$$P_*(A) + P_*(B) \leqslant P_*(A \cup B) \leqslant P_*(A) + P^*(B) \leqslant P^*(A \cup B) \leqslant P^*(A) + P^*(B).$$

From these properties we can easily show that

$$P_*(A) + P^*(\neg A) = 1.$$

Surprisingly enough, quite simple axioms on qualitative probability can be given that lead to upper and lower measures that satisfy these properties. The intuitive idea is to introduce standard events that play the role of standard scales in the measurement of weight. Examples of standard events would be the outcomes of flipping a fair coin n number of times for some fixed n. It is also easy to see using in clinical medicine a set of standard cases to play the same role.

The formal setup is as follows. The basic structures to which the axioms apply are quadruples $\langle X, \mathscr{F}, \mathscr{S}, \geqslant \rangle$, where X is a nonempty set, $\mathscr{F}$ is an algebra of subsets of X, that is, $\mathscr{F}$ is a nonempty family of subsets of X and is closed under union and complementation, $\mathscr{S}$ is a similar algebra of sets, intuitively the events that are used for standard measurements. I shall refer to the events in $\mathscr{S}$ as *standard* events S, T, etc. The relation $\geqslant$ is the familiar ordering relation on $\mathscr{F}$. I use standard abbreviations for equivalence and strict ordering in terms of the weak ordering relation. (A weak ordering is transitive and strongly connected, i.e., for any events A and B, either $A \geqslant B$ or $B \geqslant A$.)

DEFINITION. *A structure* $\mathscr{X} = \langle X, \mathscr{F}, \mathscr{S}, \geqslant \rangle$ *is a finite approximate measurement structure for beliefs if and only if* X *is a nonempty set,* $\mathscr{F}$ *and* $\mathscr{S}$ *are algebras of sets on* X, *and the following axioms are satisfied for every* A, B, *and* C *in* $\mathscr{F}$ *and every* S *and* T *in* $\mathscr{S}$:

AXIOM 1. *The relation* $\geqslant$ *is a weak ordering of* $\mathscr{F}$;
AXIOM 2. *If* $A \cap C = \emptyset$ *and* $B \cap C = \emptyset$ *then* $A \geqslant B$ *if and only if* $A \cup C \geqslant B \cup C$;
AXIOM 3. $A \geqslant \emptyset$;
AXIOM 4. $X > \emptyset$;
AXIOM 5. $\mathscr{S}$ *is a finite subset of* $\mathscr{F}$;
AXIOM 6. *If* $S \neq 0$ *then* $S > \emptyset$;
AXIOM 7. *If* $S \geqslant T$ *then there is a* V *in* $\mathscr{S}$ *such that* $S \approx T \cup V$.

From these axioms the following theorem can be proved.

THEOREM. *Let* $\mathscr{X} = \langle X, \mathscr{F}, \mathscr{S}, \geqslant \rangle$ *be a finite approximate measurement structure for beliefs. Then*

(i) *there exists a probability measure* P *on* $\mathscr{S}$ *such that for any two standard events* S *and* T

$$S \geqslant T \text{ if and only if } P(S) \geqslant P(T),$$

(ii) *the measure* P *is unique and assigns the same positive probability to each minimal event of* $\mathscr{S}$,

(iii) *if we define* P_* *and* P^* *as follows:*

(a) *for any event* A *in* $\mathscr{F}$ *equivalent to some standard event* S,

$$P_*(A) = P^*(A) = P(S),$$

(b) *for any* A *in* $\mathscr{F}$ *not equivalent to some standard event* S, *but lying in the minimal open interval* (S, S′) *for standard events* S *and* S′

$$P_*(A) = P(S) \text{ and } P^*(A) = P(S'),$$

then P_* *and* P^* *satisfy conditions* (I)–(III) *for upper and lower probabilities on* $\mathscr{F}$, *and*

(c) *if n is the number of minimal elements in* $\mathscr{S}$ *then for every* A *in* $\mathscr{F}$

$$P^*(A) - P_*(A) < \frac{1}{n}.$$

As can be seen from the theorem, by increasing the number of minimal standard elements the approximation in terms of upper and lower probabilities can be made as fine as we wish. The whole point in many clinical contexts is not to reach for too much refinement.

Within this framework we can, in a natural way, approximate Bayes' theorem or the method of maximum likelihood. An important feature of the kind of setup I have described is that it does not make sense to ask for arbitrary precision in the assignment of upper and lower probabilities to events, but only in assignment in rational numbers to the scale of the finite net of standard events. It should be emphasized that what has been developed here is not an alternative theory of statistical inference but rather a method of approximate measurement that is meant to bring a greater note of realism to clinical applications of the standard theory.

BIBLIOGRAPHY

1. Bayes, T.: 1763, 'An Essay Toward Solving a Problem in the Doctrine of Chances', *Philosophical Transactions of the Royal Society* **58**, 370–418.
2. Carden, T. S.: 1974, 'The Antibiotic Problems' (editorial), *The New Physician* **19**.
3. Gorry, G. A., and Barnett, G. O.: 1967–1968, 'Experience with a Model of Sequential Diagnosis', *Computers and Biomedical Research* **1**, 490–507.
4. Hacking, I.: 1975, *The Emergence of Probability*, Cambridge University Press, London.
5. Kunin, C. M., Tupasi, T., and Craig, W. A.: 1973, 'Use of Antibiotics: A Brief Exposition of the Problem and Some Tentative Solutions', *Annals of Internal Medicine* **79**, 555–560.
6. Peterson, O. L., Andrews, L. P., Spain, R. S., and Greenberg, B. G.: 1956, 'An Analytical Study of North Carolina General Practice', *Journal of Medical Education* **31**, 1–165.
7. Roberts, A. W., and Visconti, J. A.: 1972, 'The Rational and Irrational Use of Systemic Antimicrobial Drugs', *American Journal of Hospital Pharmacy* **29**, 828–834.
8. Scheckler, W. E., and Bennett, J. V.: 1970, 'Antibiotic Usage in Seven Community Hospitals', *Journal of the American Medical Association* **213**, 264–267.
9. Shortliffe, E. H.: 1976, *Computer-Based Medical Consultations: MYCIN*, American Elsevier Publishing Co., New York.
10. Simmons, H. E., and Stolley, P. D.: 1974, 'This is Medical Progress? – Trends and

Consequences of Antibiotic Use in the United States', *Journal of the American Medical Association* **227**, 1023–1026.
11. Stoic and Epicurean Philosophers: 1940, *Stoic and Epicurean Philosophers*, the complete extant writings of Epicureus, Epictetus, Lucretius, and Marcus Aurelius, ed. by W. J. Oakes, Random House, New York.
12. Suppes, P.: 1974, 'The Measurement of Belief', *Journal of the Royal Statistical Society* (Series B) **36**, 160–175.
13. Warner, H. R., Toronto, A. F., and Veasy, L. G.: 1964, 'Experience with Bayes' Theorem for Computer Diagnosis of Congenital Heart Disease', *Annals of the New York Academy of Science* **115**, 558–567.

MARTIN E. LEAN

SUPPES ON THE LOGIC OF CLINICAL JUDGMENT

Those knowledgeable about the subject of Professor Suppes' paper will appreciate that there are few contemporary philosophers who are equal to debate with him on this topic or to criticize his application of the various probability approaches to specific domains of empirical investigation and practice. In his present paper it is mainly in the appendix, in which he sketches his calculus for determining upper and lower probability approximations, as detailed in his 1974 paper on 'The Measurement of Belief' [1], that Suppes exhibits his special technical expertise. There he sets forth his version of Bayes' well-known theorem as a more plausible approach than other probability techniques for purposes of applying statistical methods to situations of practical belief, decision and practice. As is to be expected, his exposition provides no opportunity for critical comment.

As anyone having acquaintance with the topic knows, there are numerous types of interpretation of probability, leading to different probability theorems and calculi. Some of these are recognizably similar and more or less isomorphic, and some are appreciably diverse. Certain types of problems and applications seem naturally to call for one rather than another of the different interpretations and systems of probability calculation. And all are deficient in one or another respect – lacking in universal applicability, or leading to counter-intuitive results and even paradoxes in certain domains of application.

Of all the theories of probability, the "subjective" theories have been most subject to doubt and criticism on intuitive and philosophical grounds. Yet for the kind of problems that Suppes' paper addresses, some form of subjectivist theory does seem most appropriate. For "degree of confidence" and "confidence-value" are notions and terms that are prima facie germane to decision theory as it applies to clinical judgment in diagnosis and in therapeutic action.

In his present paper Suppes does not undertake to show the usefulness and naturalness of the Bayesian approach for this type of problem (though he does in other of his writings, and the bibliographical citations do discuss concrete cases). Others have argued, with varying degrees of plausibility and persuasiveness, that other probability interpretations and calculi can be put

H. T. Engelhardt, Jr., S. F. Spicker and B. Towers (eds.), Clinical Judgment: A Critical Appraisal, 161–166. *All Rights Reserved.*

to useful service in dealing with these same sorts of matters. Suppes does not address these arguments here, but he surely would acknowledge that it is not an either-or issue. However natural and preferably suitable he may regard the inverse or so-called "backward" probability of the Bayesian approach, he could hardly deny that other, "straightforward" statistical measures (on which, after all, any inductive confirmation or acceptance theory ultimately rests) may also be put to effective use in dealing with questions of diagnosis, prognosis, decision and practice.

Despite the subtitle and the expositional appendix, the main thrust of Suppes' remarks in the present paper is not really the comparison of the Bayesian type approach with others, nor an argument for its preferability. His paper contains very little discussion and less argument about this issue. The arguments against the direct application of Bayes' theorem to clinical practice that he cites, and his responses to these arguments, would apply equally well, *mutatis mutandis,* to *any* method of statistical inference, I suggest. Thus I read Suppes' paper as a (hortatory) discussion of the possible value for problems of clinical judgment of a probability calculus approach *at all* – i.e., without regard to which such type of approach is employed. To this I shall direct some general remarks.

It should be clear that I am in fundamental agreement with Suppes in what I take to be his real thesis. The only critical observation that I would make is that he does not seem to me to give sufficient acknowledgement (in the relevant discussion in Part II of his paper) to the possibly insuperable problem of assigning relative quantitative weights to the consequences that enter into clinical decision regarding therapeutic praxis.

So far as the purely diagnostic aspect of the logic of clinical judgment is concerned, this matter of weighting poses no special problems – at least none that cannot be dealt with by empirical statistical methods. But as Suppes rightly notes, when the decision pertains to choice of treatment, evaluation of consequences assumes dominant proportions: the probability of recovery, the probability of death, the probability of serious medical side effects, and the cost factor. To these he might well have added such considerations as the relative degree of acceptability to the patient of the possible side effects, of the financial costs, and even the acceptability of no symptomatic relief at all (with or without attendant risk) which the further option of *no treatment at all* might entail.

In his discussion of consequences as they affect clinical judgment Suppes seems to me to skate rather too lightly and quickly. He says that "the direct medical consequences and the direct financial costs of a given method of treatment . . . can be evaluated by summing across patients and ignoring more detailed individual features"; and he acknowledges that while "the unconditional or mean assessment of the consequences of particular treatments is a relatively straightforward piece of data analysis, the situation is quite different if we want to make the appropriate conditional assessment – conditional upon the variation in relevant parameters of the patient's state at the time of treatment."

Now in what he says here, and in the passages immediately following, which conclude the section, Suppes seems to me to be admitting the difficulties but ignoring their import for his program. I find it not enough for him to remark simply that "the complexity of estimating the joint probability distribution of various consequences – or symptoms – is admitted by almost everyone who has considered the problem," and to suggest only that "what seems desirable at the present time is development of an actuarial approach to both the state of health and the consequences of treatment" – an approach, he admits, that could serve only as a guideline in the treatment of individual cases.

But it is just here, obviously, that the practitioner encounters some of his most troubling decision problems. By contrast with these problems questions of straightforward medical diagnosis and prognosis are relatively simple. Whereas an actuarial approach in determining purely medical consequences, and even in estimating financial costs, can still be relevant for the individual case, even when the medical and other details of that case are ignored, the experienced practitioner will see dubious value in a statistical technique that does not relieve him of the often agonizing problem of making the "right" treatment decision. What the actuarial approach can provide him is merely a more exact quantitative estimate of what he already has a fairly good idea of on the basis of accumulated experience. The theoretical gain in predictive precision afforded by this quantitative actuarial data is in practice likely to be more than balanced by the margin for error in the conditional assessment of the unique factors, both medical and non-medical, of the individual case, with which, after all, the practitioner must deal.

Suppes speaks of recourse to the actuarial approach "for the present time." Is it really just a matter of "the present time?" Whereas in the purely medical aspects of diagnosis and prediction it may well be that increasing success will come with time in the increase of data and the refinement of statistical techniques, it is surely more difficult, if not indeed impossible, to see how informed choice of treatment is to be determined statistically without assigning appropriate weighting to the various non-medical as well as medical factors of the individual case. For surely such weighting must vary not only with the individual patient's unique medical state and financial circumstances, but also with his personal assessments and preferences. Assigning such weighting seems to me to be, if not wholly unamenable to statistical and decision theory techniques, uneconomical to the point of worthlessness.

In short, these considerations indicate that the kind of statistical techniques that Suppes is proposing must be limited to medical diagnosis and to the purely factual aspects of the predicted medical results (and possibly the probable financial costs) of alternative methods of treatment.

The preceding critical observations do not, however, detract from Suppes' main theme: that a Bayesian (or other) statistical approach could contribute greatly to accurate diagnosis, to prediction of medical results, and thus to more informed decision as to treatment alternatives.

Suppes cites a number of studies in which diagnosis has been made using Bayes' theorem with quite positive results. One such study, involving the diagnosis of congenital heart disease, "classified new patients with an accuracy close to that of experienced cardiologists" ([2], p. 148).

This notwithstanding, there is no denying that at the present time, in the present state of the art, there is no extant substitute for the clinical training and experience, the native intelligence, the "intuition," or "instinct," or skill of the seasoned practitioner, whether in physical or psychological diagnostics and therapy. At present, surely, only a beginning and unconfident practitioner would be inclined to rely on *any* kind of probability calculus or algorithmic method in substitution for his own experiential judgment. In his paper Suppes' remarks repeatedly and clearly exhibit acknowledgement of this.

For the present audience – whose temper neither Suppes nor I know – it may well be that it is unnecessary to plead the case. But we are all aware that there is a certain rather widespread resistance to the kind of program that Suppes is proposing. Thus it bears stating that the problems of clinical diagnosis and treatment should not *a priori* be regarded as beyond the capability of statistical techniques. On the other hand, as Suppes observes, it would be a mistake to demand or to suppose that such methods must or even could lead to "getting matters exactly right and organizing a body of data that will lead to certain and completely reliable conclusions about any given patient" ([2], p. 154) (What practitioner, after all, claims such infallibility using his present skills?).

On the other hand it would equally be a mistake to suppose from this that such methods could not, when developed to greater sophistication and tested empirically, lead to statistically equal or even better results for a large patient population, at least for the average practitioner. And it is an open question, one that should be decided not *a priori* but empirically, experimentally, whether such techniques might not be made to yield diagnostic and therapeutic results that are statistically equal or even superior to those obtained by the direct perceptual and intuitive judgments of even the most experienced and acknowledged expert practitioners today. (Admittedly one would be ill advised to hold one's breath while waiting!)

To those already familiar with the history of applied statistics in various domains, and especially with statistical techniques and decision theory in their more sophisticated modern development, all of the foregoing will be obvious. It is to those who, while undeniably competent and even expert as medical practitioners, are not thus familiar, that Suppes' paper and my comments are primarily addressed. But I suggest that there is another faction that must be addressed. (And here I suppose I show my sometime background as a clinical psychologist.) There are those, I am convinced, who might scoff at Suppes' proposal because of personal bias, resistance, and even nagging fear of being rendered superfluous – of one's years of schooling and experience being equalled and possibly even exceeded by unfamiliar methods and techniques – "being replaced by a computer!"

To allay this kind of fear and resistance, let me stress two points by way of concluding these remarks. First, Suppes is not proposing the *substitution* of the techniques of the statistician for the art of the practitioner. All that he is proposing is a marriage of skills. Who, rationally, can reject out of hand a proposal that may well lead to more successful diagnostic and therapeutic results?

Second, and finally, it is appropriate to remind outselves that just as no computer makes anything but a mechanical or electronic mistake, i.e., malfunctions – it is the programmer who writes the wrong program, or feeds inaccurate or inadequate data into the machine – so the role of the diagnostic and operative medical practitioner will remain dominant. It is the medical researchers and practitioners who must determine experimentally the relevant accurate medical data for the application of statistical techniques.

Though they need not be, there is no reason, of course, why the medical practitioner and the applied statistician should not be one and the same individual. All that is important is whether the diagnostic and procedural results prove to be equal or superior to those of the practitioner not employing statistical techniques. This is surely a matter that should be decided experimentally, by trial and error and modification, not by prejudice. And it is also surely not unreasonable to expect improved success from the merger of the medical arts with the methods of applied statistics and decision theory.

University of Southern California
Los Angeles, California

BIBLIOGRAPHY

1. Suppes, P.: 1974, 'The Measurement of Belief', *Journal of the Royal Statistical Society* (Series B) 36, 160–175.
2. Suppes, P.: 1979, 'The Logic of Clinical Judgment: Bayesian and other Approaches', this volume, pp. 145–159.

SECTION III

CLINICIANS ON CLINICAL JUDGMENT

EDMUND D. PELLEGRINO

THE ANATOMY OF CLINICAL JUDGMENTS

Some Notes On Right Reason And Right Action

For it is uneducated not to have an eye for when it is necessary to look for proof and when this is not necessary.

Aristotle ([4], *Metaphysics* IV, 1006a)

INTRODUCTION

Richard Cabot, one of the most celebrated of American diagnosticians, labelled his attempts to anatomize the process of differential diagnosis "a very dangerous topic – dangerous to the reputation of physicians for wisdom. . . . Physicians are naturally reluctant on such matters, slow to put their thoughts to paper, and very suspicious of any attempts to tabulate their methods of reasoning" ([8], p. 19).

Sixty years later most clinicians remain reluctant and more than a little suspicious. Many still take refuge in the "art" and its exclusive mysteries, to resist formalization of their mental operations. In recent years, the "dangerous" subject has been boldly attacked by a few clinicians with unusual facility in the language of logic and statistics so that dissection of the tangle of science, art and conjecture is now well underway ([6], pp. 17–19; [13, 15, 20, 25, 36]). They have been joined by statisticians, engineers, psychologists, and philosophers who are bringing the more sophisticated techniques of their disciplines to bear on this important and complex question.

Physicians and philosophers have puzzled for a long time over the nature of the physician's enterprise. Its dissection began with the Hippocratic physicians who first insisted on the primacy of observations of patients, refined by reason. They thus made medicine a natural science, separating it from both philosophy and religion with which it had so long been intimately associated [16].

Intuitive, hieratic, artistic and even magical elements of the physician's enterprise were not so easily removed however. Celsus, the Roman Hippocrates, spoke of medicine as an "ars coniecturalis"; William Osler, the modern Hippocrates, deemed it ". . . a science of uncertainty and an art of probability." Medicine has yet to become a science comparable in method and explanatory power with the laboratory sciences – though this is the expressed aim of some of our contemporary thinkers.

H. T. Engelhardt, Jr., S. F. Spicker and B. Towers (eds.), Clinical Judgment: A Critical Appraisal, 169–194. *All Rights Reserved.*

The puzzlement continues today in the growing tension between the scientific-actuarial and the artist-intuitionist models of clinical judgment [11]. Each view tries earnestly to understand, and thus to improve, the clinical enterprise. One seeks to transform the conjectural elements into respectable science by formal analysis in terms of probabilistic logic or decision theory. Meanwhile, the other declares the nuances of clinical judgment to be an art, insusceptible to formal analysis, and improvable only by cultivation as one cultivates painting, music or sculpture.

This essay undertakes nothing so pretentious as the resolution of what seems to be a polemically seductive polarity in viewpoints. It takes the view that: Clinical decisions can be made more rigorous logically, without reducing everything to algorithm and regression equations. While much of the process remains opaque and insusceptible to extant methods of formal analysis, this need not forever be the case.

Much of the polarization between explanations of clinical judgment seems avoidable, if two significant features which have been neglected are sufficiently taken into account. These are: First, the over-riding fact that the whole process is ordained to a specific practical end – a right action for a particular patient – and that this end must modulate each step leading to it in important ways. A value screen is thus, in a way, cast over the entire sequence. Second is the fact that no unitary explanation or logical method can encompass the several different reasoning modes and several kinds of evidence acceptable in answering the different kinds of questions the clinician must answer.

The aim of this paper is *not* to countermand the utility of scientific actuarial formulations, or offer in their stead a benignant eclecticism which allows equal place to the explicable and inexplicable, i.e., to the "science" and the "art" of clinical judgment. Rather, it hopes to locate more precisely the several reasoning modes useful at each of the sequential and simultaneous steps which eventuate ultimately in a clinical action. This localization constitutes the anatomy of clinical judgment, an essential foundation for further formalization of the process.

The complete process is a multi-step, end-oriented concatenation of decisions demanding different types of reasons and reasonings which will justify a particular course of action, for a particular patient, given that patient's particular existential situation at the time of the decision. Each step is shot through with uncertainties, some eradicable, some not. Selection of the "right" action requires optimization of these uncertainty states.

Inasmuch as the nature of the uncertainties varies, the logical instruments used to achieve optimization will vary also. Thus, the methods of deductive and inductive reasoning, of dialectic, ethics and rhetoric are each appropriate to some step. Prudent and judicious action, rather than a true statement or scientific law, is the end to which the whole is directed. No one kind of evidence is universally applicable, and no single logical or epistemic mode suffices for all.

Customarily, discussion of clinical judgment ends at the establishment of the most probable diagnosis, or at best at the selection of a treatment. The third step – whether the treatment selected ought to be instituted, how, when, under what conditions – is usually not formally analyzed. Yet, one cannot speak legitimately of clinical judgment without seeing the whole process, particularly the way the end conditions the steps that precede it.

Four questions must be examined to sustain these assertions:

(1) What is the character of the end of clinical judgment?

(2) What questions must be answered in attaining that end?

(3) How does the end project itself on each step, i.e., what kinds of reasons does it require?

(4) What are the theoretical and practical implications of the proposed view of clinical judgment?

The answers bear to some extent on the difficult problem of a general theory of medicine. Such a theory must account for the full range of medical activity – seeking general laws of disease and treatment, but also taking specific actions in the interest of specific patients. Science, art, and the virtue of prudence infuse the several operations. But medicine justifies itself uniquely as medicine – as opposed to medicine as clinical or basic science, or as the art of performing medical acts – when it ends in a decision to act for, and in behalf of, a human who seeks to be healed [28]. To the extent that it identifies features which distinguish medicine from other human activities, the analysis of clinical judgment can contribute to the emerging structure of a theory of medicine.

I. WHAT IS THE END OF CLINICAL REASONING?

When a patient consults a physician, he/she does so with one specific purpose in mind: to be healed, to be restored and made whole, i.e., to be relieved of some noxious element in his/her physical or emotional life which the patient defines as dis-ease – a distortion of his/her accustomed perception of what is a satisfactory life. Usually some event has occurred – pain, lack of appetite, fatigue, trauma, a lump, spitting blood – some sign or symptom which exceeds a certain highly personalized threshold of tolerance in such fashion as to compromise, obstruct or discolor the person's perception of him/herself and his/her health. At the point when this perception leads to the *need* to be healed, the person becomes a *patient*. This transition has two existential connotations crucial to this discussion. Becoming a patient is to become one who suffers (patior, pati), and simultaneously one to whom something is done, a recipient as distinguished from an agent (O.E.D.). The patient, then, is a suffering person who comes to the physician to have something done to assist him/her to regain his/her former state or a more optimal one.

It is important to interject here that the term "patient" does not necessarily imply a passive restoration in which the physician is the sole agent. The patient ideally also participates in the restoration. He/she is seen as bearing a burden of illness which requires some action or decision mutually arrived at for cure to take place. The patient can yield his/her moral agency to the physician only by direct mandate. Only in the rarest circumstances will the physician legitimately be the patient's moral agent [21, 27].

The end of the medical encounter, and the process of clinical judgment through which it is achieved then, is restoration and healing – some corrective, remedial or preventive action is directed at what the doctor and the patient perceive as a diminution of the patient's wholeness, each in his/her own fashion. The end is not a diagnosis, a scientific truth, testing an hypothesis or evaluating a treatment, though the knowledge derived therefrom enters into several states in making the decision to act.

The clinical decision comes at the end of a chain of deductive and inductive inferences, serially modified by recourse to "facts" and observations, usually themselves to some degree uncertain. Truth and certitude are, therefore, almost always problematic. Out of the uncertain conclusions of earlier syllogisms, a decision to act must be taken which has a different character from the conclusions that precede it. The conclusions of the earlier reasoning chains become premises for further reasoning. This is the normal condition of scientific reasoning – a cumulative and progressive series of hypotheses and conclusions, always open to further recourse to fact and experiment.

Each clinical decision is, however, a terminal and unique event in that it cannot remain forever open and it is not universalizable. It must close on the selection of one or a series of remedial actions – or none. The action chosen must be the "*right*" one for this patient. That is to say, it must be as congruent as possible with his/her particular clinical context, his/her values and his/her sense of what is "worthwhile" or "good". What the physician and the patient seek together is a judicious decision, one which optimizes as many benefits and minimizes as many risks as the situation will allow. The definition of risk is highly personal, and it turns on the patient's estimate of a danger "worth" running. The end is, therefore, not a general statement of the probabilities, but a particular statement of what a particular patient *should* do.

The criteria of a right or good decision lie not in its certitude, rigor, logical or mathematical soundness, though the probability of a judicious action is enhanced greatly when these qualities inhere in the prior conclusions on which it is built. These qualities must be secured wherever possible, but they are not sufficient for a "right" decision. They can, on the view I am propounding, be displaced or modulated by the more complex criteria of a decision "good" for this patient – what among the many things that *can* be done *ought* to be done?

Clearly, value considerations and moral issues – for both the patient and the physician – can color the selection of the facts and reasons which justify placing one diagnosis over another, the justifications accepted for taking action, and the degree to which the physician persuades or the patient assents to the final decision. The end must therefore be understood in all its fullness, because it projects itself so forcefully on the entire sequence.

The primary end of clinical judgment – a right healing action for a particular patient – imposes an atmosphere of prudence, over the whole process. Truth, for the practical intellect, is rightness with respect to human deeds, those dependent upon human will and intention. It differs thus from the truth of science which is a certain conformity with the reality it seeks to explain, and art, with the product it wishes to produce.

It is the end of medical judgment which also gives authenticity to the professions a physician makes – that he/she will use his/her knowledge and skill to "heal", according to the fullness of meaning of the word for his/her patient. Not to keep that end paramount is to make an inauthentic – indeed an immoral – profession. Medicine qua medicine is then more than a clinical or basic science applied to individual cases. It is a particularized knowledge of prudent healing actions, dependent upon scientific methods and art, but not synonymous with them.

Whatever formal analysis one favors – traditional syllogistics, the logic of probabilities, Bayesian conditional probabilities, decision analysis theory, information processing models, or judicial algorithms – each must account for the modulation and shaping imposed by the end. This would be so even if a high degree of certitude were possible at every step. Usually, however, we deal with varying "degrees of belief," and the judicious choice of actions becomes an even more critical and sensitive activity, difficult to formalize and formularize.

II. WHAT QUESTIONS MUST BE ADDRESSED AND WITH WHAT REASONING MODES?

When a person becomes a patient in the sense we have defined that state, a whole series of questions becomes crucial for him/her as a knowing and valuing being. What is wrong? Is it serious? What will it mean to me? Can it be cured, and by what means? Is the cure worthwhile? What will it cost? What *should* I do? These and corollary questions must be addressed if the process of clinical judgment is to be a complete and authentic medical judgment. They are reducible to three *generic* questions: *What can be wrong? What can be done? What should be done for this patient?* (Fig. 1) Let us look briefly at each question to see the kinds of reasoning relevant to each question, and what kinds of justifications are acceptable.

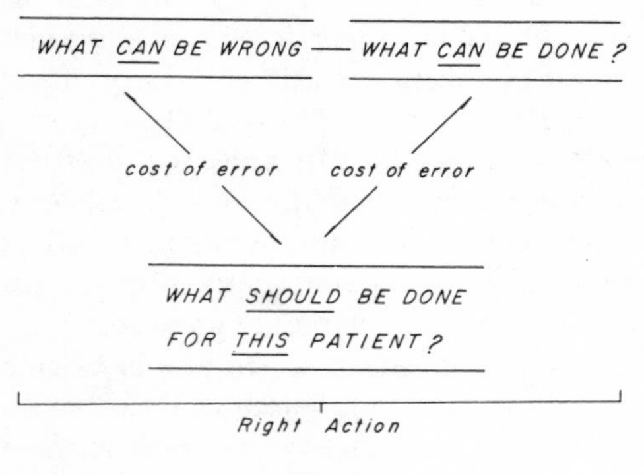

Fig. 1

A. *What can be wrong?* (Fig. 2)

This is the diagnostic and classificatory question – the first part of the process and the one which has received the most vigorous theoretical examination. Given the signs and symptoms presented by this patient, what classificatory patterns fit best? Of the possible patterns, which is most probable, and with what degree of certainty?

This part of the process most closely fits the scientific paradigm and under ideal conditions can yield a diagnostic conclusion with a high degree of

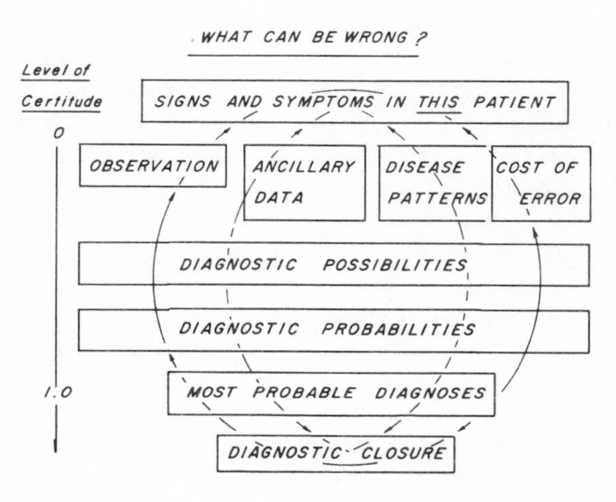

Fig. 2

certitude – i.e., the error rate approaches zero. The conditions for such certitude must be stringent, however. The input data of signs and symptoms must be reliably observed, standardized and specified; the classificatory patterns must be equally reliably determined; the probabilities of different combinations of signs and symptoms must be derived from sufficient numbers and combinations of sets and subsets of signs and symptoms; laboratory and other ancillary data must be sensitive, accurate, specific and precise. The rules of deductive or probabilistic logic must be followed rigorously.

In addition, a highly specific and sharply discriminating test is required – like a biopsy, fiber endoscopy, angiography, or an enzymatic or immunological determination. Where all these conditions are met, *diagnostic closure* can be obtained. This means that all essential criteria for a diagnosis have been met. The diagnosis could even meet the test of scientific "elegance" if it is arrived at with an economy of steps.

These rigorous conditions are only rarely satisfied in clinical reality. Bedside data are notoriously unstandardized and poorly quantified [17, 18]; laboratory tests very widely in sensitivity, specificity, reliability and accuracy [24]; "Almost all data on the general incidence and prevalence of disease are inaccurate" ([12], p. 214). Even less secure are estimates of the cost of error of a clinical act so that the hopes of decision theorists to assign a numerical value to every nodal point in a decision tree are even less secure [32, 30, 13]. Finally, probability statements say something about the population of individuals sharing some common characteristic but they are weakest in their description of the actual state of any individual in that population.

Thus, even when the rules of probabilistic logic are rigorously applied the diagnostic conclusions are still open to question. They are, more often than not, tentative "working" diagnoses and have more the quality of opinions than scientific judgments ([35], pp. 101–117). Adding more empiric data may as easily compound an error as move the conclusion closer to the asymptote of a true statement. At some point adding more tests adds only marginally to certitude if at all.

Differential diagnosis consists then in the selection of some diagnosis as more probable, or more justifying of an action than others. At some point, every diagnosis must proceed without further recourse to empirical input. The strength of the case for each diagnostic claim must be assessable and put in some order of priority against the others. Why, for example, is this chest pain due to angina pectoris and not to the fifty or more other kinds of chest pain? The selection must be made when all the empiric data are "in", and further appeals to the empiric are either impossible or insufficient to bring us any closer to diagnostic closure. This part of the differential diagnostic process is not synonymous with application of the laws of statistical probability. Indeed, a critique of the statistical argument as one among other arguments is itself requisite.

The process of differential diagnosis is really most akin to the process of classical dialectic ([2], p. 31, footnote 1 and pp. 243–247). Each diagnostic possibility can be seen as a "claim"; the strength of arguments for, and against, each claim is evaluated; each position is clarified and set against its opposite; the inherent logical probity of each is delineated. Dialectical discourse properly conducted becomes an internal soliloquy in which the clinician examines his own conclusions critically, or tests them externally against his colleagues or consultants. No new truths are discovered this way, and the dialectic ends if an appeal to experience is possible. The process is therefore not "scientific" in either the classical or modern sense of science.

Differential diagnosis is usually addressed as synonymous with an assessment of relative probabilities. Yet part of the dialectical procedure is to examine the probity of the statistical method itself. When the facts are not determinative, we must establish which diagnostic claim has the strongest logical position. This is more akin to arguing a case in court than it is to proving a scientific hypothesis. The whole effort is to make one diagnosis sufficiently more cogent than the others so that it becomes a defensible basis for decisive action.

The dialectical process is especially important in science where it has been shown that psychologically the tendency is to prove rather than disprove one hypothesis [32]. New possibilities are often uncovered only by disproving an existing hypothesis entirely. Dialectical discourse assures an adversarial stance which challenges the tendency to settle too easily for a merely "workable" hypothesis.

The "old-fashioned" clinico-pathological conference was primarily an exercise in clinical dialectic. It has fallen into disrepute, yet it remains an invaluable way to teach the process of internal soliloquizing which enables the diagnostician to examine his own reasoning critically, when the possibility of adding more observations and tests is no longer open to him. Contemporary pedagogical opinion to the contrary, the dialectical discourse of a clinico-pathological exercise more closely approximates the clinical conditions of differential diagnosis than the scientific model of serial hypothesis testing.

In answering the first of the triad of questions that make up a clinical judgment, several different kinds of reasoning, and reasons, must be employed. To understand the pathophysiology of clinical manifestations, to define and apply a classificatory schema, "scientific" reasons are appropriate. To make general statements of probability for populations, statistical logic is needed. In differential diagnosis the rules of dialectic are the most pertinent. Uncertainty pervades every step of the diagnostic process from its epistemological assumptions to its logical operations. Each of several reasoning modes is suited to optimizing different forms of uncertainty. In the next section we shall see how the practical end modifies each of these modes at every step.

B. *What can be done?* (Fig. 3)

This is the therapeutic question. Having made some decision about the nature of the patient's problem, what kinds of actions could be taken to remove or ameliorate the probable disorder?

Here again, the process is in part, and under certain stringent conditions, scientific in character. When there is a verifiable data base on the course of the untreated disease, and its modification by drugs, surgery, diet or other measure, the conditions for a scientific decision can be fulfilled. Equally indispensable are quantifiable and precise information about effectiveness and toxicity, since these must be weighed against each other before the decision is made. Under these conditions "closure" on a therapeutic action can approach certitude.

The therapeutic decision is easy to make when there is a specific and

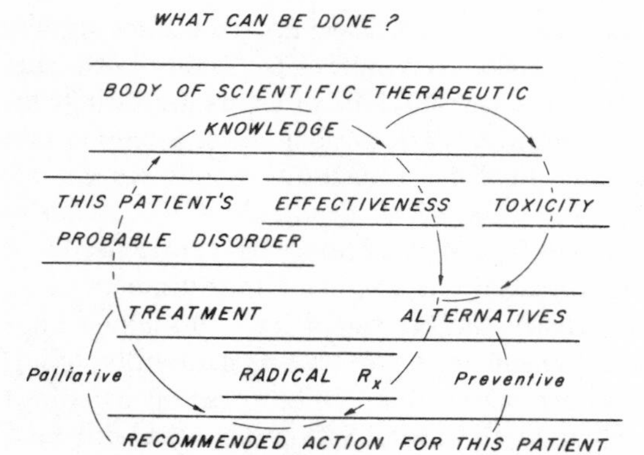

Fig. 3

highly effective treatment which demonstrably alters the natural history of the disease. Penicillin for pneumococcal pneumonia, isoniazide and PAS for tuberculous meningitis, vitamin B_{12} for pernicious anemia, or abdominal surgery for ruptured peptic ulcer are examples. In these instances, the action recommended can be scientifically validated, and it will in most instances take precedence over other considerations. The recommended procedure then becomes synonymous with right action, and little choice is open to patient or physician, though the patient may in extreme circumstances reject even this kind of advice.

Unfortunately, to a much larger degree than in diagnostic decisions, genuinely scientific information in therapeutics is scanty. The many pitfalls of therapeutic trials have been the subject of a vast literature. They are generally acknowledged to be among the most difficult experiments to design, control, carry out and interpret. Even the randomized clinical trial has come under fire recently [14]. Especially problematic are the large number of therapeutic maneuvers which are not *radical* – i.e., they do not eradicate the causal agent or the offending process. Their benefits, if any, are marginal and the ratio of risk to benefit often vacillates widely. One has only to recall the sad history of anticoagulants in myocardial infarction, the recent national dilemma about influenza vaccination, or the belated appreciation of the long-term vascular effects of oral contraceptives or antidiabetic agents.

Today, we must often decide whether to recommend complex, expensive, palliative procedures with unpleasant and dangerous side effects – as in chemotherapy for cancer. Data on effectiveness and toxicity are usually available in these circumstances, but the questions are a different order: Is the discomfort worthwhile for this patient? Is length of life more important to him or her than its quality? This kind of question must rest firmly on scientific data, but that is only the foundation for a discussion permeated by the physician's and the patient's value systems.

The therapeutic and prognostic domains may well be, as Feinstein opines, the truly unique elements of clinical science. They are also the least secure scientifically. We simply lack the long-term observations of carefully selected patients in sufficient numbers to warrant secure prognostications about the course of a particular disease in a particular patient. Without such data, it is impossible to ground effectiveness of any agent scientifically unless it is so radically effective that a few cases will suffice. If the untreated mortality is 100%, any change will indicate effectiveness, as it does in treating subacute bacterial endocarditis or tuberculous meningitis with specific antibiotics.

The choice of what action to recommend involves more questions of value, by far, than diagnosis. The closer we come to the end of the process of clinical judgment – the right action – the less useful and less available is the scientific model. Reasoning becomes, in smaller part, scientific and probabilistic, and in larger part, dialectical – arguing one alternative against another without recourse to new factual data.

C. What should be done for this patient? (Fig. 4)

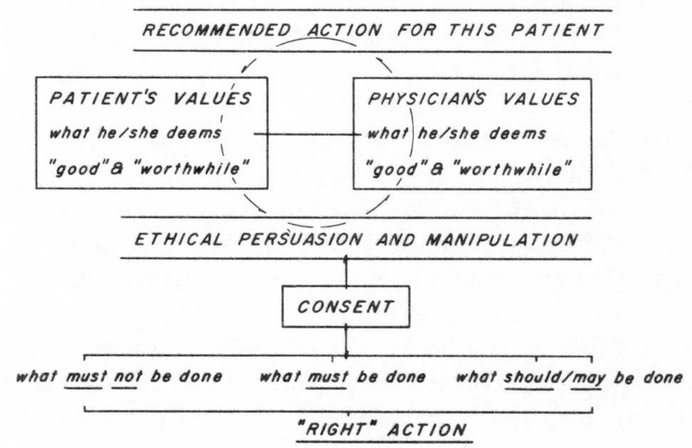

Fig. 4

Having decided what the probable diagnosis is, and what treatment can be expected to be most effective and least harmful, the final question in clinical judgment is whether the treatment *should* be used in this patient, and what alternatives can be offered. The right action – the best one for a given patient – is not always synonymous with the logically or scientifically deduced action. The amount and kind of information needed to secure a diagnosis may be quite different from that needed for decisive action. It is the task of the last stage of clinical judgment to make these distinctions with as much precision as possible.

Decisive action frequently involves the counterposition of what is good scientifically, what the physician *thinks* is good, and what the patient will accept as good. Scientific, personal and professional values intersect each other and can be in conflict. The scientific evidence and the probability statements about diagnosis, prognosis and treatment become arguments for or against a choice of alternatives – to do nothing, or, if something is to be done, what must be done, what may be done, and what should be done.

If the patient exhibits the minimal criteria for an acute abdomen, signs of intracranial pressure, a stab wound of the heart, he/she *must* be operated upon; with severe blood loss, bacterial endocarditis, tuberculous meningitis, or diabetic coma proper, medical therapy *must* be instituted. On the other hand, a non-strangulated hernia, symptomatic gall stones, or mildly symptomatic benign prostatic hypertrophy are things that *should be* treated but may not be in the *must* category. Finally, hemorrhoids, varicose veins or lipomata *may be* removed surgically; or mild hypertension, a cold or sinusitis *may be* treated, but need not be.

When it comes to making the *right* decision – the *judicious* one for this patient – the categories of *must not, must, should* and *may* can all shift, depending upon a myriad of factors in the patient's life situation and his/her notion of what he/she deems worthwhile. In the more obvious situation – like a Jehovah's Witness rejecting transfusion or a Christian Scientist rejecting surgery – the shift is to the *must not* category regardless of the canons of good science. In the more usual case, the categories into which a recommendation falls may shift several times in the course of the illness.

For one person, a lipoma, acne or sebaceous cyst is something to be ignored; for another, it is a horrid blemish, so emotionally disabling that it demands removal. For this person *may* becomes *must.* For an elderly man or woman chemotherapy of disseminated cancer may mean a few months more of life, less pleasantly lived. For this patient, a *should* or *may* becomes a *must not.* For another person, the few extra months to do some essential things may make the rigor of treatment worthwhile. For still another, mild depression, anxiety or headache is tolerable, and can be coped with without resort to daily medications; to another, as the twenty-four million Americans daily taking mood modifiers will attest, even slight anxiety must be treated ([9], p. 141).

In making the "right" decision for an individual patient then, personal, social, economic, and psychological characteristics of the patient must be factored in. They clearly may modify, or even nullify, the scientifically cogent or logically consistent answers to the questions of what can be wrong? And what can be done? Here, where we are closest to the end, the *telos*, of the whole process, scientific modes of reasoning and scientific reasons are least pertinent, and indeed must be submitted to drastic revision in consideration of the patient's value choices.

The reasoning at this stage is mainly dialectical, ethical, and rhetorical. Physician and patient together must clarify the relationship of one recommendation with its opposite and weigh the reasons for each action. The ethical nature of the discourse at this point is obvious. Conflicts in the obligation of patient and physician to each other must be resolved before a "right" action can be settled upon. Where these conflicts are fundamental, the medical relationship may even be severed by either party, to be resumed with another physician.

Once the ethical and logical possibility and "strength" of arguments for one action over another are assessed by dialectic, then the reasoning becomes "rhetorical" – in the more classical sense of artful persuasion, of relating a dialectically established decision to prudent action, generating belief of another kind from scientific or logical cogency, belief that this particular action should be taken in preference to all others ([34], especially pp. 27–28).

We enter here the delicate realms of how much persuasion is ethically defensible, how vigorously the physician should pursue his/her own priority system. Does he/she believe the scientifically recommended action must take priority? Is he/she a therapeutic enthusiast who prefers to "do something" rather than nothing? Does he/she think people should be "strong" and ignore minor infirmity?

Likewise, the patient can use artful persuasion to modify the physician's scientific or dialectically secured recommendation to gain an action more congenial to his/her view of what is good. The patient's vulnerability can easily tip the resolution of conflict in the physician's favor. But these are subtle influences, as well, which can significantly shape the physician's final judgment.

The last question in the sequence then – what should be done? – the capstone question, as it were, which completes the whole structure, is the most prickly. Scientific and semi-scientific conclusions of varying degrees of certitude are examined under a light strongly tinged with moral hues. The accessibility of the questions to scientific modes of reasoning declines, as does the degree of certitude, as we move from determining what *is* wrong. to what *can* be done, to what *should* be done. The optimization of several kinds of uncertainty remains a central concern even when the conclusions are scientifically defensible.

The intermingling of several modes of reasoning – clinical scientific, probabilistic, dialectical, rhetorical and ethical – does not suggest immunity from explicit statement nor deprecation of attempts to secure truly scientific conclusions. But, it is warrant for setting limits on the utility of any of the current explanatory models – scientific, probabilistic, Bayesian, judicial simulation, decision analysis, inference, information or catastrophe theories. Perhaps the fullest explication will come from some combination of these new theories and more classical modes of reasoning.

III. PROJECTION OF THE END OF RIGHT ACTION ON THE PROCESS OF CLINICAL JUDGMENT

We have seen that clinical judgment requires confrontation with three generic questions which end in a particular decisive action for a particular patient. Several interdependent reasoning modes are requisite and suitable to each question, and no one question is entirely isolable from the others. We are ready now to turn briefly to a more specific delineation of how the end of right action projects itself upon the way the questions are answered. We might accordingly label this section "clinical prudence" – or, medicine viewed as a virtue.

Whenever possible, the clinician will try to achieve diagnostic closure, i.e., satisfy all the stringent criteria for certitude in locating his/her patient precisely in a classificatory schema. This is the admirable aim of scientific medicine; it is the ideal taught in academic settings; the more it is achieved, the better foundation there will be for the decision to act, at least some uncertainty will be eliminated.

Many clinical realities can upset the orderly processes of diagnostic reasoning: the urgency of the patient's condition, the absence of a diagnostic set or algorithm which fits the presenting signs and symptoms closely enough, lack of specific tests to discriminate among probable diagnoses, incomplete or inconclusive probability estimates, limitations on data selection related to cost or geographic locale, unreliability of the patient as historian, and the dilemma of weighting the probability of a serious vs. nonserious, and treatable vs. nontreatable disorder, when none of these possibilities is definitely excludable.

Experienced clinicians deal with these commonly occurring inevitabilities with interpretative schemata, which enable them to act prudently on inadequate information and incomplete evidence. These schemata can be loosely regarded as rules of clinical prudence, empirically derived but certainly not mysterious sources of inspiration. We can examine a few of these. (Fig. 5)

SOME RULES FOR CLINICAL PRUDENCE

1. ACT TO OPTIMIZE AS MANY BENEFITS, MINIMIZE AS MANY RISKS AS POSSIBLE.

2. THE SERIOUS AND TREATABLE MUST NOT BE MISSED; THE NON-SERIOUS AND NON-TREATABLE MAY BE MISSED.

3. USE THE CLINICAL OCKHAM'S RAZOR: DON'T MULTIPLY CAUSES, DISEASES, TESTS OR TREATMENTS WITHOUT JUSTIFIABLE NECESSITY.

4. "REST" THE CASE FOR ANY DIAGNOSIS OR TREATMENT WITH RELUCTANCE.

5. CLINICAL SKEPTICISM IS THE ONLY GUARD AGAINST THE TYRANNY OF THE "ESTABLISHED" DIAGNOSIS, OR ANCILLARY DATA, AND THE FINDINGS OF COLLEAGUES (LAB, X-RAY, ETC.).

6. MAINTAIN A HIGH INDEX OF SUSPICION FOR UNCOMMON MANIFESTATIONS OF THE COMMON.

7. "HOOFBEATS DON'T MEAN ZEBRAS," UNLESS ZEBRAS ARE IN THE VICINITY.

8. WHEN THE DATA ARE "IN," CONTINUING DEBATE IS THE SAFEGUARD AGAINST ERROR.

9. RECOGNIZE YOUR OWN CLINICAL STYLE, PREJUDICES AND BELIEFS ABOUT WHAT IS "GOOD" FOR PATIENTS.

10. BE WARY OF HUNCHES, INTUITIONS, E.S.P.--GAMBLE WITH YOUR OWN FATE, NOT THE PATIENT'S.

Fig. 5

For one thing, the clinician thinks not so much of meeting the criteria for diagnostic closure, but how much evidence is "enough" to take an optimizing action. Diagnostic closure for example in streptococcal pharyngitis might demand sore throat, plus cervical adenopathy, tonsilar exudate, and positive culture for Beta hemolytic streptococcus. The physician is justified in giving specific treatment with penicillin on any combination of these signs and symptoms and positive culture. But he/she usually acts presumptively – taking a culture, treating, and awaiting the culture to determine whether to continue or discontinue penicillin. If the patient, however, lives in a family or school in which streptococcal illness is frequent or currently present, the most minimal clinical signs may justify treatment even without culture.

Acute abdominal pain is another example. It may result from a wide variety of conditions all of which can present similar signs and symptoms. Examples are: ruptured peptic ulcer, appendicitis, twisted ovarian cyst, strangulated internal hernia, mesenteric adenitis, pancreatitis, cholecystitis. Some of these conditions require immediate surgery, some delayed surgery, some none at all, and in some surgery is positively dangerous. The surgeon looks not for definitive diagnosis but for the least signs of acute abdomen – spasm and rebound – which justify laparotomy. Only in this way can he/she assure that the treatable lesion is not missed. Each surgeon is allowed a certain number of normal appendices or negative explorations, provided he/she has *no* cases in which lifesaving surgery is missed.

In every branch of medicine, there are disorders for which specific and dramatic treatment or genuine cures exist. They must not be missed. They shape the diagnostic process very often out of proportion to their probability in a given case or in the general population. Some are rare like pheochromocytoma; some relatively rare like parathyroid adenoma; some are rare only in a particular locale like malaria or actinomycosis or blastomycosis. still others were once common, but now infrequent, though still important like syphilis, malaria or diphtheria.

The clinician's attention therefore may not be focused on getting all the information needed to fill out the most probable classificatory set in its entirety. Rather, more emphasis is placed on the information he needs to take decisive action, or to "rule out" the conditions which in the interest of the patient's safety must not be missed. The clinician must find justification for his/her actions, therefore, in terms of what is right and prudent for the patient. If a serious treatable disease cannot be adequately "ruled out," therapeutic trial may be used as a diagnostic maneuver, i.e., specific treatment is undertaken for the serious disorder, and diagnosis arrived at indirectly by observing the effect. Such a course of action may be taken justifiably even if the probability of a less serious, nontreatable disease is much higher.

A corollary principle, which to some extent balances the search for the treatable and serious, is the admonition "don't look for zebras when you hear hoofbeats." This is a precaution against accepting the rare and esoteric even if the symptom complex is suggestive. Obviously, one cannot justify missing the more common, or losing the more unusual manifestations of a common disorder in the search for the curable but esoteric. Of course, on the African plain, hoofbeats do mean zebras and not palomino ponies or quarterhorses.

Equally influential is the clinician's Ockham's razor – "plurality of causes and diseases is not to be assumed without necessity." Clinicians, particularly in the clinicopathological conference, are conditioned to seek unitary explanations of signs and symptoms. This is another of the organizing principles which become part of the projective schema used by clinicians to select, screen and weigh data and to justify their decision. The same advice is salutary in therapeutics – the rule of therapeutic parsimony states that treatments must not be multiplied without necessity.

Another rule of prudence is the cultivation of a "high index of suspicion" – another quasi-rational way to cope with uncertainties and avoid "missing" something. Clinicians who follow this rule are unusually sensitive to the most minimal criteria for a diagnosis. They use that sensitivity to open up previously unsuspected possibilities or challenge what appears to be a more firmly established diagnosis.

Indeed, the most dangerous ground for clinical judgment is often not in making a new diagnosis, but in maintaining an index of suspicion high enough to challenge what appears to be an established diagnosis. Physicians, like all humans, tend to rest their cases if no contrary position challenges it. They become locked into or even enamored of the patient's classification, which can render them unresponsive to the most obvious signals of a new disorder, or an original diagnostic mistake. The more prestigious the original diagnostician or institution and/or the more respect one has for the colleague who has made the diagnosis, the more seductive is this error.

Skepticism is, as Santayana so mordantly said, the "chastity of the intellect" – and an indispensable attribute of the prudent clinician is an untiring skepticism about every aspect of a diagnosis or treatment. Perhaps the greatest failing of the contemporary clinician is an overadulation for laboratory and x-ray data fed in by specialists, whose techniques grow more arcane and also more potent daily. Far too often, a right action is justified by the uncritical acceptance of a chemical determination, a biopsy or an x-ray examination.

I do not intend to provide here a complete primer of clinical prudence. These few principles derived from experience illustrate how the intensity of the clinician's concentration on the end – a good decision for his/her patient – provides a projective schema which shapes every step on his way [1].

Projective schemata are, of course, individualized and personalized. They summate what a particular physician holds to be "good" medicine. Which rules of thumb are selected, why, and how, are fascinating questions still very much open to logical and psychological examination. A whole series of attitudes of mind, logical and epistemological assumptions, are combined to make up a clinician's diagnostic and therapeutic "style," just as genuinely as the way word and phrase are used to make up a writer's style. The clinician's "style" is really a statement of the reasons he/she will accept as justification for a particular action, for setting aside certain probabilities, or for earnestly persuading a patient to take one course rather than another, equally earnestly propounded by his/her colleagues.

Are these not the elements of the "art" of medicine, and are they not closed to precise analysis, the way the style of Horace or Camus is closed? I think not. I agree with Claude Bernard's comment in warning against sole reliance on clinical instinct and clinical sense.

> Everyone knows in fact that habit may give a kind of empirical knowledge of things sufficient to guide practitioners, even though they may not always be able to account for it at first. But what I blame is willfully staying in this empirical state and not trying to get out of it. ([5], p. 203)

Each clinician's rules of prudence need more precise statement. Many are held in common by all good clinicians. Their utility or fallacy can and must be better studied. Most theoreticians have concentrated on the *normative* aspects of clinical judgment. We need to know more about the descriptive nature of the process – who uses what justifications, and how do they benefit or imperil the outcome for the patient? How much difference in mortality, disability, side effects, costs is there between a good and a poor decision maker?

Some physicians are therapeutic enthusiasts. They treat more often than not; they require less justification for using a medication than parsimonious therapists, who insist on demonstrable evidence of effectiveness and frown on the use of placebos or chemical coping for minor disorders. How much of "style" is simply peer conformity, personal intolerance of ambiguity, an obsessive compulsion to "prove" the diagnosis, or satisfy some idealized notion of what constitutes clinical "science"? Much of the cost of medical care, and even its outcome and satisfaction for the patient can turn on how the clinician plays the "game" of clinical judgment.

They psychological components in different reasoning styles are coming under closer investigation [33, 11]. We also need more extensive empirical descriptions of the reasoning habits of "good" and "bad" clinicians. What is the effect on outcome of different habits of reasoning? Which ones are crucial, which incidental and which positively deleterious? In this era of concern for the quality of care, answers to these questions cannot be avoided for very long.

There remain, to be sure, certain features of clinical judgment which for the moment seem closed to explicit analysis. Clinicians have their undeniable moments of sudden insight and discovery. Some clinicians are more consistently accurate in their insights than others. What does happen at that precise moment when the clinician decides that there is "enough" information to "make" a diagnosis or "take" a decisive action? Or, when a set of signs and symptoms suddenly "assembles" itself into a new classificatory pattern? Or, when a patient's clinical features suddenly "fit" a previously described pattern; or, when the precise datum which will answer a puzzling question is suddenly identified?

There is no reason to suspect that clinical diagnosis is unique in this regard. The phenomena of "insight" and "discovery" are common to scientific and nonscientific endeavors. Polanyi and Lonergan have attempted to describe the deep epistemological structures and relationships which permit the leap from the obvious and the known to the previously unknown. Polanyi holds that scientific knowledge is never wholly explicit but resides in tacit understanding which provides the belief which makes thinking about a truth possible. Lonergan links belief to immanently generated knowledge. Polanyi's "discovery" and Lonergan's "insight" are likened by their authors to the "Eureka" of Archimedes [19].

To what extent these epistemological notions can account for the obscure features of clinical reasoning is problematic. What must be avoided is the easy appeal to some special illumination peculiar to clinicians. To resort to terms like "art" or intuition is to impede explication of a socially significant process. Whatever name we use to subsume the indefinable elements in the process, the effort to explicate them further is a moral as well as intellectual responsibility.

This warning against subjectivism is consistent with a major thesis of this essay – that the nature of the end projects itself on the steps leading up to it. Everything in medicine ultimately is judged by its end – the healing of a patient – even those steps we think most immune to pre-logical or subjective interpretation like pattern definition and recognition. In defining a disease pattern, for example, we must choose among an infinite set of properties – signs, symptoms, tests, etc. That resultant pattern is useful only if it includes a finite, and manageable number of characteristics. In making the finite selection we apply a test of utility, related to the purpose of the pattern definition. That purpose is making a "right" decision for a particular patient.

In pattern recognition, the selection is similarly shaped by the end. We seek sufficient congruence with the pattern characteristic of the more serious or the more treatable disorder. We select in favor of patterns which optimize the decision for the patient even though they may be less defensible by the rules of logic or probability.

Statisticians, logicians and engineers can help the clinician to apply criteria of validity to the characters he/she selects and to manipulate the sets once they are defined. But these aids are on limited use without definition of the purpose for which the classificatory set is to be used. Sokal, a pioneer in mathematical taxonomy, recognizes that in medicine we may limit ourselves to fewer characters than might be taxonomically desirable because we want those most pertinent to decisive action for a patient [31].

Projective schemata shaped to the practical end of decision making are to a significant degree subjective. But they do not provide warrant for a wholly subjectivist interpretation of clinical judgments. The schemata can themselves become objects of critical scrutiny. We can in fact often judge whether a clinical diagnosis, treatment or decision is the right one by the outcome for the patient. The test of reality is always there in medicine, and it sharpens and corrects the projective schema, and the values assigned in each of the more formal inferential steps. It is as Ernan McMullin said of the interpretive schemata of science which contemporary philosophers of science have adduced to counter the ultra-objectivism of logical positivism:

> But what must be emphasized is that the skills of interpretation that help bridge the gap between the formalism of inference and the subjectivism of sheer personal assertion are themselves continually responsive to the demands of the objective order [23].

In medicine we have more immediate and dramatic verificatory possibilities than in many other sciences (death, recovery, disability, complication, dissatisfaction). We can aim realistically therefore to reduce to a minimum the residuum of indefinable elements in each step of the process of arriving at clinical action. This applies even to the more subtle nuances in style which distinguish superior from merely average clinicians.

IV. SOME THEORETICAL AND PRACTICAL IMPLICATIONS

Clinical judgment is what physicians do that most clearly distinguishes their enterprise from other human activities. If there is to be a theory of medicine therefore, an understanding of clinical judgment must be a central element – though certainly not the whole of such a theory.

This essay has argued two major theses which bear on a theory of medicine: that clinical judgment is justified and defined by its end – a decisive action for a human in distress and that different kinds of reasons, and reasonings, are appropriate to the formulation of a "right" end.

There are several important implications for a theory of medicine or at least its prolegomenon as well as medical education. Only two of these will be examined briefly: 1) Is medicine a science, art, or virtue or all three; 2) What information should a physician have?

(a) Medicine – science, art or virtue?

Sober has, in this conference, summarized the arguments on both sides of the art-science controversy and they need no repetition [32]. He firmly identifies clinical reasoning with scientific reasoning. On the view we have proposed there is no question that significant parts of the process do fit both the classical and modern descriptions of a science.

Deductive, inductive and retroductive inference are used to evaluate, interpart and explain clinical observations. Hypotheses are tested by further observation, experiment and measurement, and reliability, accuracy and standardization of data are sought, if not always achieved. These criteria of the scientific method are most clearly pursued in making the diagnosis and selecting and evaluating treatments. Medicine is even more convincingly a science when it seeks explanations of clinical phenomena, in theories and mechanisms of disease. Medicine then fits Einstein's definition of science as:

> . . . the century-old endeavour to bring together by means of systematic thought, the perceptible phenomena of this world into as thoroughgoing an association as possible. To put it boldly, it is the attempt at the posterior reconstruction of existence by the process of conceptualization. ([10], p. 24)

But the principal aim of medicine, its distinguishing feature, is not the explanation of clinical phenomena, useful as this may be in raising the clinician's enterprise above empiricism and technicism. Explanations in Einstein's sense are the proper business of the sciences basic to medicine, but they are not synonymous with medicine. Medicine exists as medicine only when it engages in the full range of activities which constitute clinical judgment and which lead to decisive action in the interest of a particular patient.

In its complete expression as medicine, the clinician's enterprise embraces non-scientific as well as scientific reasons and justifications. This essay alludes to the dialectical nature of differential diagnosis, the ethical construction in deciding what is "good" for this patient, and the rhetorical dimensions in the mutual persuasion essential to a decision taken jointly by patient and physician. These latter are skills traditionally imparted by the liberal arts and humanities. They may be exercised in medicine on data validated by scientific method. While dependent upon the reliability of these data, medicine is not identical with the means whereby they are validated.

Is medicine then in any sense an art? Not as the term is used in the current art-science controversy. We have argued that clinical judgments must not be assigned to realms of the intuitive and ineffable which find their source in a presumed special intellectual light possessed by clinicians. I would prefer to reserve the word art in medicine to the perfection of the things done by the physician – the craftsmanship without which the decisions taken would be improperly, unsafely or clumsily done. The art of medicine lies in the degree of perfection each clinician exhibits in history taking, physical examination, performance of manipulative techniques like surgery and various diagnostic maneuvers – the work done.

These must not be mindless activities. They are a tekné in the classical sense – knowing what to do, how to do it, and why one does it. Art and tekné were synonymous in Aristotle and both involve reasoning, ". . . art is identical with a state of capacity to make, involving a true course of reasoning" ([4], *Met.* VI, 4, 1140a, 10). Aquinas, extending the Aristotelian notion called art a recta ratio factibilium, thus distinguishing it from science, a recta ratio speculabilium and prudence, a recta ratio agibilium. The undeviating determination of the work to be done (art) is distinguished from the undeviating determination of what is to be known (science) and the undeviating determination of the act to be done (prudence) ([22], p. 7, [3]).

If we carry this distinction a trifle further medicine becomes not only science and art, but also the virtue of practical wisdom – ". . . a true and reasoned state of capacity to act with regard to the things that are good or bad for man" ([4], *Met* VI, 5, 1140b, 5–6). Aquinas expanded the Aristotelian concept somewhat. He argued that, while art gives the capacity to reproduce a good work it does not assure that the product will be used for a good end. That remains to the virtue of prudence – "Consequently, prudence which is right reason about things to be done, requires that a man be rightly disposed with regards to ends" [3].

In this sense then, medicine is both art and virtue. It is art since it is certainly concerned with the perfection of what it does – with the skills needed to make logically secure clinical decisions and to do the procedures decided upon. But it is also a virtue since it must make "right" choices about the ends and purposes for which the decisions and actions are produced. Medicine must not only perform well but act well, it must choose what should be done to heal a particular patient whose good is the true end of the whole activity.

Medicine is then all three – science, art and virtue synergistically and integrally united in the clinician's daily activities. To disarticulate one member of this triad from the others is to dismember medicine – the essential feature of which is the special relationship each holds to the other. When this happens one becomes a scientist, an artist or a practitioner, but not a physician. Clearly, any unitarian formalization of the clinician's activity is bound to be misleading, and to define a part of what constitutes medicine but not the whole of it.

(b) *What should a physician know?*

I have explored elsewhere the many meanings of the term medical humanism – which can so easily become a slogan for whatever version of intellectual or practical activities we wish to justify [26]. Here it is only necessary to reiterate that medical humanism is both a cognitive and compassionate response to the person of the patient. It embraces humanism as a set of cognitive skills largely derived from the humanities and these are integral to the conception of medicine as science, art and virtue. Humane medicine – indeed moral medicine – requires that the physician understand the distinctions between these intellectual and practical activities, the kinds of reasons each may adduce, their limitations when applied to each other's realms, the different sources of their methodology and the different subject matter appropriate to each.

If there is any merit in these proposals, they should be reflected in what it is physicians are expected to know. If medicine is a science, and art, and a virtue (in the classical sense) – medical students will need a more explicit education in the non-scientific components of clinical judgment. The skills required for sound dialectical, ethical and rhetorical reasoning modes must be more explicitly incorporated into professional education – especially in the clinical contexts within which decisions for patients are being made. The liberal arts, and the humanities, even though they might have been presented in pre-medical studies, need vigorous refurbishment in medical education where it will become more obvious to the student that he *needs* the liberal attitudes of mind to function fully as a physician.

The liberal arts have a legitimate place in medicine, not as gentle accoutrements and genteel embellishments of the medical "art", or even to make the physician an educated man. Rather, they are as essential to fulfilling the clinician's responsibility for prudent and right decisions as the skills and knowledge of the sciences basic to medicine.

These are very good reasons for reinfusing the liberal arts into medical education in addition to, but not necessarily to replace, the liberal education in the universities. Premedical education is shallowly rooted for a variety of reasons: The pressure for admission to medical school converts non-scientific subjects into obstacles to be endured rather than essentials to an educated life. Similarly, the humanities may have become too specialized themselves to fulfill their functions as instruments of the liberal arts. Appreciation of the humanities may come more slowly than in the sciences – usually only after the young physician has had some experience of the moral nature of so many of his decisions. Then, the intellectual constraints of a life dedicated only to professional studies are only belatedly perceived. For all these reasons, the liberal arts and humanities are today being taught as part of a medical education in a significant number of schools [29].

The full implications of the anatomy of clinical judgment for the philosophy of medicine cannot be examined further in this essay. Clearly a theory of medicine must account for this, the physician's most characteristic activity. Hippocrates was partially right to make medicine a natural science; Celsus to call it a conjectural art, and Osler, a science of uncertainty. Any theory of medicine must accommodate all these elements and in the special way demanded by the need to make particular decisions for particular patients. A theory of right thinking, right doing and right acting must then be related to each other. The theory of medicine must also be shaped by the end to which all medical activity points – a right healing action – for that is what defines medicine uniquely.

> It is a simple fact but almost universally ignored in modern thought that when one loses sight of the end of one's thought and action, the thought and action waver between fanaticism and futility. ([7], p. 174)

Yale University
New Haven, Connecticut

BIBLIOGRAPHY

1. Abercrombie, M. L. J.: 1974, *The Anatomy of Judgment*, Penguin Books, London.
2. Adler, M. J.: 1927, *Dialectic*, Harcourt, Brace & Co., New York, p. 243, 247.
3. Aquinas, T., *Summa Theologiae*, transl. by W. P. Hughes, Leonine text, 1a 2ae, Question 57, art, 3–4.
4. Aristotle: 1940, *Metaphysics*, transl. by W. D. Ross, The Clarendon Press, Oxford.
5. Bernard, C.: 1927, *An Introduction to the Study of. Experimental Medicine*, Macmillan, New York, p. 203.
6. Black, D. A. K.: 1968, *The Logic of Medicine*, Oliver & Boyd, Edinburgh.
7. Buchanan, S.: 1938, *The Doctrine of Signatures: A Defense of Theory in Medicine*, Keegan, Paul, London, p. 174.
8. Cabot, R.: 1916, *Differential Diagnosis*, W. B. Saunders Co., Philadelphia, vol. 1, p. 19.
9. Chambers, E. E., Inciardi, J. A., Siegal, H. A.: 1975, *Chemical Coping: A Report on Legal Drug Use in the U.S.A.*, Spectrum Publ., New York, p. 141.
10. Einstein, A.: 1974, *Out of My Later Years*, Citadel Press, Secaucus, New Jersey, p. 24.
11. Elstein, A. S.: 1976, 'Clinical Judgment: Psychological Research and Medical Practice,' *Science* **194**, 696–700.
12. Feinstein, A. R.: 1967, *Clinical Judgment*, Williams Wilkins, Baltimore, Maryland, p. 214.
13. Feinstein, A. R.: forthcoming, 'The Haze of Bayes, the Aerial Palaces of Decision Analysis, and the Computerized Ouija Board,' *Clinical Bio-statistics* **39.**
14. Feinstein, A. R., and Wells, C. K.: forthcoming, 'Randomized Trials vs. Historical Controls: The Scientific Plagues of Both Houses,' Association of American Physicians, Washington, D. C.
15. Jacquez, J. A. (ed.): 1964, *The Diagnostic Process*, University of Michigan Press, Ann Arbor, Michigan.
16. Jaeger, W.: 1944, *Paideia: The Ideals of Greek Culture,* transl. by G. Highet, Oxford, New York, III, 17–19.
17. Koran, L. J.: 1975, 'The Reliability of Clinical Methods, Data and Judgments, first of two parts,' *New England Journal of Medicine* **293,** No. 13, 642–646.
18. Koran, L. M.: 1975, 'The Reliability of Clinical Methods, Data and Judgments, second of two parts,' *New England Journal of Medicine* **293**, No. 14, 695–701.
19. Kroger, J.: 1977, 'Polanyi and Lonergan on Scientific Method,' *Philosophy Today*, Spring, 2–20.
20. Ledley, R. S.: and Lusted, L. B.: 1959, 'Reasoning Foundation of Medical Diagnosis,' *Science* **130**, 9–21.
21. MacIntyre, A.: 1977, 'Patients as Agents,' *Philosophical Medical Ethics*, ed. by S. Spicker and H. Engelhardt, Reidel, Dordrecht, Holland, pp. 197–212.
22. Maritain, J.: 1949, *Art and Scholasticism With Other Essays,* Charles Scribner & Sons, New York, p. 7, notes 4 and 5.
23. McMullin, E.: 1974, 'Two Faces of Science,' *Review of Metaphysics* **27**, No. 4, 655–676.
24. McNeil, B. J., Keeler, E., Adelstein, S. J.: 1975, 'Primer on Certain Elements of Medical Decision-Making,' *New England Journal of Medicine* **293**, No. 5, 211–215.

25. Murphy, E. A.: 1976, *The Logic of Medicine*, Johns Hopkins Press, Baltimore Maryland.
26. Pellegrino, E. D.: 1974, 'Educating the Humanist Physician: An Ancient Ideal Reconsidered,' *Journal of the American Medical Association* **227**, No. 11, 1288–1294.
27. Pellegrino, E. D.: 1977, 'Moral Agency and Professional Ethics: Some Notes on Transformation of the Physician-Patient Encounter,' commentary on Alasdair MacIntyre, 'Patients as Agents,' in S. Spicker and H. T. Engelhardt (eds.), *Philosophical Medical Ethics*, D. Reidel, Dordrecht, pp. 213–220.
28. Pellegrino, E. D.: 1976, 'Philosophy of Medicine: Problematic and Potential, *Journal of Medicine and Philosophy* **1**, No. 1, 5–31.
29. Pellegrino, E. D.: 1976, 'Preface' in T. K. McElhinney (ed.), *Human Values Teaching Programs for Health Professionals*, Society for Health and Human Values, Philadelphia, pp. ix–x.
30. Raiffa, H.: 1968, *Decision Analysis: Introductory Lectures on Choices Under Uncertainty*, Addison-Wesley, Reading.
31. Schwartz, W. B., Gorry, G. A., Kassirer, J. P., and Essig, A.: 1973, 'Decision Analysis and Clinical Judgment,' *The American Journal of Medicine* **55**, 459–472.
32. Sober, E.: 1978, 'Learning Theory and Clinical Practice,' in this volume, pp. 29–44.
33. Sokal, R. R.: 1964, 'Numerical Taxonomy and Disease Classification' in J. A. Jacquez (ed.), *The Diagnostic Process*, University of Michigan Press, Ann Arbor, pp. 51–68.
34. Tversky, A.: 1972, 'Elimination by Aspects: A Theory of Choice,' *Psychological Review* **79**, No. 4, 281–299.
35. Weaver, R.: 1968, *The Ethics of Rhetoric*, Henry Regnery Co., Chicago, Ch. 2, pp. 27–28;
36. Wulff, H. R., 1976, *Rational Diagnosis and Treatment*, Blackwell Scientific Publications, Oxford, England.

MARJORIE GRENE

COMMENTS ON PELLEGRINO'S 'ANATOMY OF CLINICAL JUDGMENTS'

Dr. Pellegrino's statement of the end of clinical judgment, and of the questions that need to be asked in the course of the physician's execution of his task in making such judgments, appears to me unexceptionable. And it is of course both correct and important to distinguish between the concrete, time-bound, practical and humane goal of clinical practice on the one hand and the more generalizing, temporally less pressing, theoretical and less personal context of scientific inquiry on the other. As a philosopher interested in the conceptual structure of science, however, who cherishes a hunch that the conceptual analysis of clinical judgment would reveal a structure not so *very* different – despite the clear difference in ends – from that of scientific thought, I find a good deal to question in the categories Dr. Pellegrino uses in his exposition. He seems to be distinguishing "science, dialectic, ethics, art and rhetoric" in the spirit of the Aristotelian division of the sciences, a division, in my view, almost wholly inappropriate, in the terms in which Aristotle made it, to any modern philosophical discussion of the nature of science or disciplines like clinical medicine that exist in some practical, if not in a strict conceptual, relation to it.

Consider the chief characteristics of Aristotelian as compared with modern science. An Aristotelian science depends for its possibility 1) on the existence of a strictly delimited subject genus, in respect to which 2) we can make fairly simple inductions to necessary and commensurately universal first principles themselves intelligible 3) by a special faculty of *nous* ("intuitive reason") in such a way that 4) external, literally true, necessary conclusions follow from them – conclusions amounting to a body of knowledge radically distinct from matters of mere judgment or opinion. Only deductions from such unique, necessary first principles constitute knowledge. Dialectic, by contrast, is valid reasoning from non-necessary premises; it can always be used, as Aristotle shows at length in the *Topics*, on either side of an argument, but it never produces knowledge, only opinion. Art or *techne*, for Aristotle, is making, aimed, not at knowledge, but at the production of something: a building, a poem, possibly also health. It is this definition of *techne*, also, apparently, from which Dr. Pellegrino wishes to dissociate clinical judgment. Clinical practice in general is perhaps an Aristotelian *techne*, but the

H. T. Engelhardt, Jr., S. F. Spicker and B. Towers (eds.), Clinical Judgment: A Critical Appraisal, 195–197.

judgment of this physician at this bedside today is not an art – though it does seem to be the exercise of an art, like, say this lyre-player's rendition of a hymn to Apollo on the occasion of this festival. (I think that probably Dr. Pellegrino wants to resist the affiliation of clinical judgment to *techne* because many who speak of that kind of skillful performance as "art not science" are thinking of it as "intuitive" and so not "rational" – and that is indeed a character foreign to Aristotelian art (or craft), also, one hopes, to clinical judgment, which need not be irrational simply because it is not wholly mechanical. Ethics, further, is the discipline that articulates ultimate ends; it is governed by *phronesis*, practical wisdom, which of course does enter into the practice of clinical medicine, concerned as it must be with the choice of right actions. And rhetoric, the art of persuasion, also has its role to play in medical practice, as it had much more massively in Aristotle's time.

What concerns me most, however, is modern science; not a set of necessary universal propositions distinct in kind from opinion, but a set of relatively well established rationally and empirically justified but always defeasible beliefs. This differs from Aristotelian science on every count. 1) Short of the (impossible) realization of a unified science, with no divisions of subject matter, it is nevertheless the case that the subject-matters of our sciences constantly change, expand and overlap with one another. The biological sciences in particular are what C.F.A. Pantin called "unrestricted" sciences, borrowing from other disciplines any information or techniques that seem appropriate to assist them in their work. They cannot by any stretch of the imagination qualify as subject matters for Aristotelian science. 2) Induction for modern science, moreover, *always consists* in the movement from "inadequate information and incomplete evidence" ([1], p. 183) to tentative hypotheses, subject to further revision after further testing. The data are infinite, our assertions finite. The resultant gap cannot ever, on principle, be closed. True, the clinician does indeed have to decide on such inadequate evidence more fatefully than does the experimental scientist. Yet, apart from the difference in their goals, their conceptual situations are precisely parallel. 3) Thirdly, there *is* no *nous*. Since the world we live in is infinite and open, not finite and closed like Aristotle's, there is no way for us to get an ultimate grasp on first principles literally stateable for a given fixed region of the real. 4) And, fourthly, therefore, the conclusions we draw from our premises, however confidently we may rely on them, retain at bottom the logical uncertainty inherent in all our claims to orient ourselves in our world. If we make deductions, they had better be valid ones: neither scientists nor clinicians want to reason *badly*. But beyond pure deduction, both scientists and clinicians have to rely on clues they have interiorized from experience and/or authoritative teachers in order to focus on the puzzling situations they are trying to interpret. Granted that their aims, and therefore the urgency as well as the human relevance of their decisions, differ, still, the conceptual structure of their thinking and of the way they ascribe reasons and learn, increasingly, to make more reliable judgments: these are, it seems to me, very strikingly parallel. Dr. Pellegrino refers in passing to what he calls Polanyi's concept of "discovery" as a "Eureka" experience. What is most significant about Polanyi's account of tacit knowing, however, is his description of the complex, temporally extended structure of knowledge as relying-on-clues-in-order-to-attend-to-what-they-are-clues-to. Perception, practical skills (including that of the experienced clinician), the practice of theoretical science and, indeed, of aesthetic judgment, all share, if in significantly different ways, this common structure. If philosophers of science would pay closer heed to the character of clinical medicine, with its massive judgmental component, they might, I believe, learn much that too exclusive an attention to the most abstract and theoretical branches of science has allowed them to overlook. That is not to say that clinical judgment should not be made as "scientific," that is, as precise, as possible, in the sense that it should use whatever objective techniques it can find for its improvement and its support. But what is most important philosophically, in my view, is the opposite insight: that precision comes to be, and is maintained and increased, only under the auspices of judgment, of human-all-too-human efforts to make sense of situations that always outrun our control.

University of California at Davis
Davis, California

BIBLIOGRAPHY

1. Pellegrino, E.D.: 1979, 'The Anatomy of Clinical Judgments: Some Notes on Right Reason and Right Action', this volume, pp. 169–194.

ERIC J. CASSELL

THE SUBJECTIVE IN CLINICAL JUDGMENT*

INTRODUCTION

Someone telephones the doctor that he has had increasing dull pain in the right side of his abdomen and back for several hours. While not exactly nauseated, he is repelled by the thought of food. He thinks the pain is the same as that of his wife when she had her gallbladder attack. The patient's complaint is clearly subjective and of the type that most often initiates the medical act. Yet, in its subjectivity, the report lies in a domain of medical practice that is least understood, or more precisely that is least systematized. The deficiencies of medical practice in regard to the subjective are highlighted by the increasing use of the problem-oriented medical record as a tool of medical education. In *Medical Records, Medical Education, and Patient Care; The Problem-Oriented Record as a Basic Tool,* Lawrence Weed [4] points out the equal importance of patients' subjective experience with objective, measurable facts of medicine. Further, Dr. Weed gives excellent examples of the kind of profiles of patients' personal and social lives that should be included in any complete medical record.

It has been my experience, however, that the problem-oriented medical record *as actually used* by the medical students and house officers at the New York Hospital (many of whom trained at other schools) is a sterile instrument that rarely meets the goals set for it precisely because it is deficient in its recording of the subjective. It has also been my experience that many younger (and too many older) physicians distrust the patient's report of his own symptoms and experience.

One more comment seems necessary to document the problem of the subjective in medical practice. The social medicine movement of earlier decades succeeded in establishing the importance of the patient's social and personal history as part of any complete medical record. And yet, a generation later that aspect of a patient's history is usually confined to personal habits (tobacco, alcohol, etc.), job history and some marital and family facts. The movement was successful in getting some aspects of the patient as subject into his own care but in a fashion often uselessly sparse because only the objective facts of the patient's existence are recorded.

H. T. Engelhardt, Jr., S. F. Spicker and B. Towers (eds.), Clinical Judgment: A Critical Appraisal, 199–215.

THE SUBJECTIVE IN MEDICINE[1]

I believe the subjective involves four aspects: the sociologic person, the unconscious, experiencer, and assigner of understandings. The sociologic person includes the patient's past and present cultural set, his roles (attorney, mechanic, father, son, etc.), the unfolding story of his life and his important others. The subjective also includes the unconscious self, generally considered to be conflicts, repressed materials, drives and motivations – matters not volitionally available to the conscious mind. Then there are the things experienced by the body or person of the patient. These may be objectively non-confirmable as in the instance of abdominal pain or objectively confirmable as in the instance of fever. In all instances, after the experience has passed it is non-confirmable, although subsequent objective evidence may allow it to be inferred. Lastly the subjective includes the meanings assigned by the patient to the experience and events he reports. This includes feelings evoked, beliefs about the nature of disease or illness as well as its causes, and extends to feelings, perceptions and beliefs about physicians and the world of medicine.

Obviously these four aspects of the subjective in medicine are not truly separate and cannot be made to remain separate except for purposes of explication, as I am doing here, and except for action as when the physician asks questions. Together they are the subject *in* medicine and the subject *of* medicine. They are, in other words, the patient.

Indeed it would be difficult, for example, to distinguish that part of the patient that assigns meanings from, say, the unconscious, or even from the perceiving experiencer. But I hope to show that these domains of the subjective can be kept apart, if only in action, well enough to serve the physician as he makes his diagnosis and treats his patient.

THE SOCIOLOGIC PERSON

Two domains of the subjectivity of the patient have received the most attention in medicine: the sociologic person and the unconscious. There is ample evidence that the sociologic person is important in medicine. Diseases as disparate as tuberculosis and coronary heart disease are influenced in their occurrence by the life history of the person who has them. The malnourished black child from a large ghetto family has a much higher probability of acquiring tuberculosis than the white suburban child of a Bell Telephone supervisor, especially if there is an old person with tuberculosis in the crowded ghetto apartment. And the Bell Telephone supervisor has a greater probability of dying of a myocardial infarction than the unemployed father of the black child, especially if the supervisor smokes cigarettes, is sedentary and has a family history of coronary heart disease. Modern medicine has made much of the contribution of sociological variables to disease production (although even Virchow was politically radicalized by his awareness of the contribution of social factors to typhus prevalence in Silesia). But while probabilities may be crucial to directing the diagnostic thinking of a physician, probabilities are often not as necessary in making a diagnosis as what the sick person *says*. More important, however, is that probabilities are objective parameters of the sociologic person, not subjective. The probabilistic person is seen as propelled towards his expected pattern of disease by facts of his existence which he (usually) did not create and over which he (often) has no control. In other words it is not that the sociologic person cannot be construed as a part of the subjective domain of medicine – part of the domain of the consciousness of the patient, but rather that its current use by medicine is primarily in the objective domain: measurable or at least objectively confirmable personal facts linked to confirmable facts about the lungs or the coronary arteries. Even Weed's use of the subjective in the profiles of patient examples is aimed towards useful objective information. This kind of information might help in the interpretation of other diagnostic information (in the sense that "divorced women with young children are often . . ." and so on). But such data, while undoubtedly helpful and even often vital in clinical decision making, are part of the subjective of the physician more than of the patient. Sociologic parameters do not tell us about *this* divorced woman, but rather about the class of divorced women. In a sense the use of these facts to speak for the person so much occupies the physician's head (not to say his or her preconceptions) that they prevent the physician from hearing what *this* divorced woman has to say.

THE UNCONSCIOUS

The other aspect of the subjective that has received widespread attention in medical practice is the unconscious. It is fair to say that one of Freud's major contributions to medical practice was to put the person as subject back into medicine. It is not necessary to detail the evidence to support the belief that there are unconscious determinants of symptoms and disease. The widespread acceptance of psychosomatic medicine is an acceptance of the influence of the mind on the body. Oddly, it is conceived of as the influence of the *unconscious* mind on the body: a mind not available to the volition of the person and almost always conceived of as causing illness or symptoms. It is somewhat interesting, if only historically, that this view maintains the Cartesian duality but gives the (separate) mind some control over the body. This view of the duality is not nearly as meaningful or useful as the view of mind and body as a polar duality discussed by Guttentag [1]. Much as North and South cannot exist without each other and mutually influence each other, so also do mind and body. Of more immediate importance to our discussion is that the way the unconscious domain of the subjective is most often used in medical practice excludes the subject. That is to say, the patient is seen as not knowing what his unconscious contains or is doing. It is not under the patient's control and even the patient's words are an untrustworthy guide to its contents and its actions. We, the physicians, take it that we know better than the patient what are the unconscious determinants of his symptoms or his disease. While that may well be the case, it is an uncharitable view of the patient since that view bypasses the consciousness of the patient just as effectively as does the dominant view of the sociological person.

To view the unconscious as solely an inaccessible domain with only negative effects on the body is also, I believe, an uncharitable view of the unconscious. This is not the place (nor am I the person) to attempt a summary of the various views of unconscious process. Nonetheless, it seems important to point out that any such summary would have to include the understanding that the patient's unconscious is part of his subjective aspect that is able to communicate to another person. And, the unconscious and its communication are legitimate expressions of that person. Unconscious communications are not merely unsuspected leaks of inadequately repressed material but purposeful expressions. As the unconscious can speak, it can also be spoken to [2]. Finally, and of more direct importance to medicine, the unconscious appears able to communicate with the body. These phenomena which can be glimpsed in biofeedback techniques, hypnosis and certain yoga feats suggest a relationship of the subjective person to the body that is at present minimally understood.

Thus, just as the sociologic person is often used and viewed in medicine as an objective thing apart from the consciousness of the patients, so too is the unconscious used and viewed. For both sociologic person and unconscious, such understandings tend to diminish the potential importance to medicine of those aspects as factors of subjectivity.

EXPERIENCER AND ASSIGNER OF UNDERSTANDINGS

I believe I am warranted, if only for methodological reasons, in making a distinction between the experiencer and the assigner of understandings within the subject. When the patient reported the dull pain in the right side of his abdomen and back, which he thought was similar to his wife's gallbladder attacks, he was actually reporting two distinctly different things: first, the *experience* of a body sensation; and, second, his *understandings* – the meaning attached to the body sensations. The assignment of understandings was on at least two levels; that the sensation was painful, and that it might be gallbladder disease. Considerable attention has been paid to the "subjectivity" of reports of pain, (for which, the way it is usually used, you may substitute the word "unreliability"). It has been shown that what patients say, or that their behavior in response to pain, is influenced by their ethnic group [5]. But that is merely another way of saying that the report of pain generally includes some kind of meaning to the patient. On the other hand, I do not know of studies which show that what some people call "pain," others label "itch" or "tickle." And that seems to be because what people call pain are sensations arising from a discrete group of nerve fibers called pain fibers. What is shown by differing reports of what is presumed to be the same kind of pain, is that valuation by the patient is part of the report. A seemingly different kind of the assignment of understanding is that which implies that the pain is from a gallbladder attack. Indeed, many patients when reporting their illnesses never tell of symptoms but simply use diagnostic terms. "First I got a cold and then it turned into bronchitis." Or, "last week I had a virus and then it started my colitis off."

The dominant voice of the patient with abdominal pain was the voice of the assigner of understandings (the pain is gallbladder disease) rather than the experiencer. And it is common to say that the patient's understanding might not be an accurate reflection of the meaning of his symptom. But meaning in whose terms? It must indeed be an accurate reflection of the meaning of the symptom to the patient (unless he is lying – the unusual case). Rather when we speak of the patient's understandings being inaccurate, we mean inaccurate by the doctor's standards. To dismiss the importance of the patient's understanding is to dismiss him as subject. After all, whatever the actual disease, its importance to the patient, even if it is fatal, will depend on how it meshes back into the subjective – the collectivity of his meanings. Thus, to deal adequately with the subject of the patient we must find within the patient's report of an experience two things: what is the matter in our terms, and who is the patient. Since the patient's report of an experience is so heavily influenced by his understandings, careful questioning about the report of the experience offers the opportunity of finding out about *both* the disease and the patient.

To understand the process of the assignment of understanding it may be useful to view it in the same manner that has been used to relate a speaker to what he says [3]. A speaker, in relating a sentence, is saying something about something; the speaker is thus the *coupler* between *what* he says, and what he says it *about*. Thus when the patient tells us about a symptom, he does the same thing. He is the *coupler* between the experience and his verbal report of the experience. As we have already seen, to assign language to the report is to assign meaning, and to give meaning is to give his understanding of the experience.

But it is obvious that the report of experience is not the experience itself and that the memories and sensations that constituted the original experience remain available for reinterpretation. This is not to deny that perception is to some degree guided by the assignment of meaning, and that what steps the patient takes to broaden his perception of an event – for example, taking his temperature, feeling his abdomen, trying to remember what was eaten and so forth – are also guided by the assignment of understanding. But, if the raw sense data of an experience remain available for re-interpretation, than a new coupler – the questioning doctor – can be introduced between what is *said* (the assignment of meaning), and what it is said *about* (the raw sense data of the experience). To put it another way, the doctor can insert himself between the patient as experiencer and the patient as assigner of understandings. This may seem a complicated way of saying that to obtain the story of an illness, the doctor asks the patient questions, but I think more is being said. Since by eliciting answers the physician is forced to continue to use the patient as coupler (someone saying something about something), how does he disengage the patient as assigner of understandings? The classic answer is that the physician offers (at least to himself) an alternative hypothesis. That is to say, if the doctor believes the man with abdominal pain does have gallbladder disease, he asks questions about the pain, its relationship to food, previous episodes, associated symptoms, and so forth, until he has enough evidence to support his tentative diagnosis. Or if such evidence is not forthcoming, he rejects that diagnosis, proposes (to himself) another diagnosis, and repeats the procedure. In doing this, he is using the same reasoning that will guide his physical diagnosis, choice of diagnostic tests or other diagnostic aids. If he is a good physician, he will use strict standards for hypothesis testing, and if he is a poor physician, he will use loose standards.

The classic picture I have just provided of the physician as *history taker* is inadequate for several reasons. First, the physician will only find what he already knows. Each case may flesh out his conception of the disease as the patient describes somewhat different expressions of symptoms, but he cannot find that for which he has no diagnosis. Second, he will miss in the patient's symptoms the unfolding process that the disease represents within the patient. Thus he will be held to basically static or structural views of disease rather than as a process occurring through time. After all, the patient, as the container of the memory of succeeding events influenced by the abnormality within him, is virtually the only source of information about such a process (reinforced by the physician's observation of the change in physical findings and laboratory tests over time). Third, in that classical method of questioning, the doctor will emerge knowing little about the person who has the disease, the subject of his work. The idiosyncratic differences presented by different patients are seen by some physicians as *obstructions* to the diagnostic process rather than as an inherent part of the diagnosis. Finally, the doctor will not learn how the person of the patient interacts with basic pathologic mechanisms to produce the illness that is *this* person with *this* disease.

I would like to expand somewhat on those four points in a somewhat different vein. The classic understanding of the physician taking a history from the patient by extracting those, admittedly subjective, facts of the illness in order to make a diagnosis of disease is deficient because it excludes the person from the diagnosis. It is by now cliche in medicine to speak of the importance of the "person of the patient" to diagnosis and treatment. We do not label as cliche the saying "a coin lesion of the lung is a carcinoma unless proven otherwise." Why is the former a cliche but the latter not? Because "treat the patient as a person" is more a wish than agreed upon necessity and does not rest on established facts and procedures which lead to a defined course of action. It is more an ethical imperative than a medical directive. I would like to point out, however, that the clinical expression of the primary diseases of our time, arteriosclerotic heart disease, hypertension, diabetes, degenerative joint disease, and perhaps the malignancies depend on the individuating characteristics of the patient. Further, in marked contradistinction to the infectious diseases or the surgical diseases, the patient is the primary agent of his own treatment. That is to say, the patient must change his life style or behavior and comply with often complicated treatment requirements. This was brought home to me the other day as I tried to discuss with a Chinese-speaking patient the use of insulin, diet, and exercise for the treatment of diabetes through an inadequate interpreter.

Contrary to directives based on analogy to the infectious diseases, what is needed are directives for obtaining information about the patient as person now necessary because of the change in the disease pattern and the preponderance of chronic disease. This involves some understanding, in methodological terms, of how understandings are assigned by the patient to the sense data of experience – the meaning of experience. An experience is given meaning along four planes, or conversely in reporting sense data, the patient modifies it to conform to his understanding along four planes. They are: a) the space-time continuum – when and where things happened; b) the assignment of value in relationship to other values of the self – the process of adjectivization; c) the assignment of causality – where did the thing come from and what is it expected to do; and finally, d) the relationship of the experience to previous experiences of the self or of significant others.

When patients feel pain, or experience fever, diarrhea and abdominal pain, they interpret it to themselves along those four planes. I believe that virtually no alien body sensation can go uninterpreted although the degree and complexity of the interpretation may vary widely – as in dismissing a twinge or conversely building a case for cancer on the twinge. What I am describing is a process in which the original body sensations can lead to interpretation which is followed by a search for other data in memory or at the moment that aids in, confirms or disconfirms the interpretation. In this process some sense information will be dismissed as irrelevant, and other sense data perhaps will be elevated in importance within the patient. The assigner of understanding (within the patient) takes a history from the experiencer (within the patient) in much the same manner that the physician questions the patient.

The importance of examining the various dimensions along which the experience is given meaning is not only that by so doing the physician will obtain a more accurate picture of the original experience, but also a more accurate understanding of the subjectivity of the patient.

THE SPACE-TIME CONTINUUM

People seem to vary enormously in their ability to report details about the time and place (including place on the body) of experiences. In part, this appears to be idiosyncratic – some remembering in great detail events of the distant past and some assigning everything before yesterday to the hazy past. It is my impression that the future may be handled by different individuals in a manner similar to their handling of the past. Time and place are often related . . . "It must have been 1975 because that was the summer we were in

New Hampshire." Time may be used as a mechanism of denial so that the events become difficult to recall because they cannot be located in time. In an opposite manner, "I remember it as though it were yesterday" becomes a means of maintaining the largeness of the event. Some handlings of time do not seem to be due to individual variation. Terribly threatening events that are remembered seem to remain, for most people, close in time – "that accident couldn't have been a year already." Similarly, a far future but bad prognosis is often handled as though the threat were immediate. In all of this time seems to be dealt with as though it were a spatial not a temporal dimension. The more important the event is considered the closer to the present self it is perceived. An understanding of this is important to the physician both in obtaining the history of an illness and in discussing prognosis, future therapies and so forth. A patient with a right-sided abdominal mass and fever was considered to have an ameboma because his symptoms started while on an Asian journey. In fact, however, the first episode of abdominal pain had started a month earlier in London but was attributed by the patient to carrying a heavy suitcase (to borrow from a later topic). Careful questioning that related times and places on the trip to body sensations produced the classic story of an appendiceal abcess. The original appendicitis, the relief of pain with its rupture, a few days of fever, then quiescence, followed by a return of fever, pain, and a perceived mass.

From what we know of the patient's subjective use of the time-space dimension in relating a story, we may be able to tailor our description of his future as we think it will happen. The basic point is that an objective outline of the future temporal relationships of a disease process will be necessarily reprocessed by the subjectivity of the patient in terms of his past use of that dimension. And that reprocessing influences the course of the patient's illness. For example, one patient with rheumatic heart disease, told that an operation will probably be required in two years, lives in constant and continuing dread of the surgery, behaving as though it is imminent. For another patient, saying that the medication he is taking now is to forestall an event that might occur many years in the future (as in the prevention of stroke or heart disease in hypertension) is the equivalent of removing any sense of the importance of the event and therefore of taking the medication. Thus, the objective time scale of a disease or a projected course of action provides *merely the language for time, only the patient's subjectivity can provide its meaning to the patient* and the meaning is what the patient will use to direct his actions and his fears. As with all the dimensions of the subjective, the person's subjective sense of time-space can change. Indeed, some changes in time sense from childhood to old age appear developmental but events in one's life also can produce change in an individual manner. "What my illness taught me is that I have to value each day."

THE ASSIGNMENT OF VALUE

A second plane of subjectivity is the assignment of value to experience, or the process of adjectivization. Adjectives used by the inner assigner of understanding do not merely modify nouns such as for example pain, swelling, dizziness, or nausea, rather they give those experiences a value relative to other values of the self. A pain may be described as mild, or excruciating, uncomfortable or sore. One patient describes the pain as merely discomfort while for another patient the same kind of pain is unbearable. The assignment of value is both idiosyncratic and held in common. Certain kinds of pain, for example, pressure on a nerve root, will be described with the same kinds of adjectives by most people and this commonality allows adjectives to enter the diagnostic process much like an objective referent. On the other hand, the experience of the pain may be adjectivized differently. In this context, I can point out that physical behaviors such as writhing or grimacing, or conversely sitting stoically, add to the process of value assignment. For some people experiences are either black or white – things are terrible or simply fantastic. For others, things are more varied but the colors of their adjectives are always intense. Every experience, bad or good, is assigned a strong term. For others everything is subdued, nothing is fantastic, nothing is terrible. Pain may be awful but never terrible. Their weakened leg is not "useless" but "rather awkward." Their whole, near fatal illness was "a bore." The ethnic or cultural contribution to the assignment of value is well known.

The experiences of life enrich adjective use because these are values assigned relative to the other values of self. Since it is virtually impossible to speak without using adjectives, speakers quickly display their palette of value terms. On the other hand, many phrases or utterances do not include adjectives but their voice pitch, inflection and modulation emphasize or deemphasize the words, serving the same valuative function as the adjective in displaying the weight the speaker assigns to the words or thought. Consequently, as the physician obtains the story of an illness, he has available to him the scale of value assignment used by the patient. Consciously or not, those adjective usages and speech inflections (as well as clothes and body motion) allow him not only to understand the patient's value scale but also to revalue in his own terms the experience being reported. Similarly, since values are assigned relative to other values of the self, the physician has the opportunity to find out how important a symptom or possible consequence of an illness may be. Mild diminution in vision may be well tolerated by a patient who does not like to read but the same loss would be "intolerable" to an avid reader.

The pattern of valuings of a person is a basic constituent of that person. When the patient complains that something is "terrible" it is useless for the physician to say that it is not terrible. For the patient the thing is "terrible", "useless", "hopeless", "sad", "wonderful", "amusing", "interesting", "dirty", "smelly", etc. It is part of their understanding of the thing. And as it is part of their understanding of the thing, it is part of them as a person for we are constituted by our meanings and our meanings include the values we assign. For the physician to argue with his patient in order to change the valuing of something from "terrible", for example, to "mild" is not to change a word but the person himself. When two people look at a plate of six raw oysters one sees six oysters revolting and the other six oysters delicious. Induced to eat those oysters, a person may change his valuing of oysters from "revolting" to "delicious" because he has changed in regard to oysters. (The six remains a constant, however, and that is the infinite advantage of the number in the realm of value.) The patterning of value may change with experience and with that change comes a new understanding of experience and an alteration in the person himself.

This is an appropriate place to point out again that dividing the assignment of understandings into four planes and even separating these dimensions from the sociologic person or the unconscious is a useful procedural device for a physician but does not correspond to the person as he lives and thinks.

The person is a functional unity and these planes and dimensions interpenetrate. The assignment of value is heavily influenced by the sociologic person, by the cause of a symptom as perceived by the person (pain that is heart disease versus pain from a sore muscle), as well as by repressed unconscious material and even by the inner relationship of person and body. That these levels are a nexus rather than discrete does not detract, however, from finding out how a person assigns value. On the contrary, understanding how a patient assigns value is an important clue not only to the complete entity but to how the interpenetration occurs.

THE ASSIGNMENT OF CAUSE

A third plane of subjectivity is the assignment of cause to the sense data of experience. No event can be experienced without a search for cause. I have called this the dimension of causality, rather than cause and effect, because as every event is understood in part by understanding its cause, it is also given meaning by a conception of what will follow from the event, in other words, what it will subsequently cause. As is the case with space-time and the patterning of value, understanding the cause of one individual event does not occur apart from the person's whole pattern of causal understandings. To believe that his pneumonia was caused by a bacterium or virus, the patient must conceive of micro-organisms as existing and as being part of the general class of causes. If that understanding exists, it may be applied by the patient to a whole range of phenomena experienced in his body. Thus, alien body sensations may be subsumed under their cause as in "I had a virus last week." The physician also has such a causal nexus but his understandings may be different. To him virus may mean for example coxsackie, varicella-zoster, or mumps virus in their most specific technical sense as well as a more general and vague sense he shares with the patient of viruses as a cause. This distinction should allow me to make clearer what I mean by cause as part of subjectivity. The mumps virus is in the physician's understandings, an objective technical thing having among other specific characteristics physical shape, and also a distribution among the population or in the tissues of the host. By that general class of cause called "a virus" no such specific virus is meant, although specific viruses remain the objective referent on which the subjective is based. Rather, the subjective allows experience to be classified and put at rest, so to speak. Causes within subjectivity such as "viruses" always have antecedent causes such as "I was run down", which may have its own antecedent such as "I was under a lot of strain" and so forth. Viruses, in subjectivity, are not considered serious and so they usually are not seen as causing other serious events. The important thing for this discussion is to realize that when the patient says "I had a virus last week", he is saying very little about the sense data of the experience but a great deal about his causal nexus. For the information to be meaningful in classic diagnostic terms, the doctor must inquire of the experiencer within the patient for the specific symptoms to which the understanding "virus" was applied. To accept that the word "virus" means to the patient what it means to the doctor may lead to serious diagnostic errors, as may acceptance of the patient's words "bronchitis", "sinusitis", "stomach upset", or what have you. (Those words contain not only cause but also value assignment and time-space information since they are often more specifically inclusive than "virus".) However, not to listen to those words and to hear only the specific sought after symptoms is also to miss valuable information about the person. The causal relationships tell not only how the patient views this illness, but how all illness may be perceived in terms of cause. The train of antecedents I suggested before may ultimately have led to some psychological conception of disease causality. In another person the ultimate nexus may be self-blame and sin. The first words of a patient of mine who was found to have a cancer of the breast were, "I knew it, I'm being punished." As it is useless to argue with people about their time-sense or valuations, it is equally useless to tell them that carcinoma of the breast is not punishment for illicit behavior. Rather, it is important to see the disease and its treatment within the causal network of the patient. And that network is revealed in the words of the patient when symptoms are related. The patient's conception of cause may have to be sought but to understand the subjectivity of the person (indeed the person himself) that search is as vital as the hunt for symptoms.

THE ASSOCIATION OF EXPERIENCE

The fourth plane helpful in assessing the subjectivity of the patient is that in which an experience is given meaning by its association with the experiences of significant others. The patient has noted pain and stiffness in his fingers. The cause is not readily apparent – that is, he cannot remember doing something to the hand that might cause pain. His knowledge of the word includes the conception of arthritis, but, at age forty, he seems too young. His mother had arthritis, and remembering her, he concludes he also has arthritis. The experience of the mother appears to be dealt with as though it is part of his experience. Many other examples could be given to show the influence on the patient's assignment of understanding that is exerted by the experience of family. That a young woman with an elevated cholesterol interprets every pain in her chest as angina may seem reasonable to a physician since many of her relatives died at an early age of coronary heart disease. We find it reasonable because we know the relationship between familial hypercholesterolemia and premature coronary heart disease. While the patient may use those objective facts to support her fears, the fear is part of her subjective experience of her parents. For example, another person with normal blood pressure and headache worries always about stroke and we are not surprised to find that a parent died of a stroke. The fear of the patient that he has the disease of the parent is believed by many to arise from unconscious determinants such as unresolved oedipal conflict, or unresolved guilt. Such interpretations may be correct, but whatever the reason, patients behave in a manner suggesting that the experience of significant others is part of their own experience when they assign understanding to their own experience. Further, I am frequently surprised at how difficult it is to dislodge the patient's belief that he has the same problem as the parent even in the face of considerable counter-evidence. As the physical therapist was working successfully on her previously painful knees, the patient was sure the therapy would be useless because her mother had been crippled by arthritis. Two different lines of thought are suggested by these observations. The first is that subjectivity, the collectivity of meanings, extends beyond the physical confines of the person. Such a conclusion, which is neither new nor world shaking, achieves its importance in this context because when the doctor talks to the patient he generally acts as though he is facing a unique free standing individual. The physician does not see the continuity of the patient with the parents or siblings and neither may the patient. The continuity might not even be important if it did not so color the assignment of meaning to experience. But more striking, *that continuity tends to determine the experience.* It is in this area that the relationship of the unconscious domain of subjectivity to the assignment of understanding seems most obvious. The relationship to parents and other important people does tend to occupy the content of repressed or conflictual material making much of that content unavailable (except by inference), to even the most skilled questioner in the limited time available to the physician taking a history.

On the other hand much of the continuity of experience is available to the patient on reflection and so also is the patient's belief that the experience of significant others may determine his own experience even if he does not consciously know why that should be the case.

Where the physician obtains a family history he generally obtains facts of family disease, such as diabetes, as well as who is alive or dead and their ages and so forth. That the mother may have had diabetes influences the probability that the patient has diabetes but does not help in the diagnosis. On the other hand the fact that the mother had diabetes may have a great influence on how the patient interprets and reports increased urination or other symptoms of diabetes. It may have even more influence on how the patient behaves after he is told of his own diabetes. I should like to stress that this frequent clinical observation should not be confused with interpretations as to its psychological cause. The subjectivity of the patient influences not only the assignment and report of the meanings of an experience but the patient's interpretation of subsequent events in the illness. It cannot be overstated that the interpretations of the patient help determine the way the disease will be expressed in the patient and the patient's behavior towards the disease. Not only the symptoms but also the totality of meanings, and actions that follow these meanings, are the illness.

Another line of thought follows from these observations. Patients do not only borrow from the experience of family but also from that of friends and associates as they give understanding to their own experience. Physicians are sometimes frustrated to discover that the misinformation of friends or even casual acquaintances is given more weight by the patient than the doctor's knowledge. Why should this be? It seems to me that uncertainty is the primary impetus for the assigner of understandings. Uncertainty is intolerable. Where the previous experience or knowledge of the person is inadequate to give meaning to the events, others must be consulted. The fact that opinion from the unknowledgeable is influential cannot be because they lack knowledge, but rather because they have knowledge. But what knowledge can that be? I suspect that the associates have two advantages over the physician in the service of certainty. First, they *know cause* in the same terms as the patient, and second, the patient thinks they *know him* because they are like him. That is to say that the experience of the associates is useful in giving meaning to the patient's experience because the associates and the patient are similar. As one's subjectivity includes parents and siblings, it also includes the surrounding group. In providing their experience or knowledge the members of the group do two things: they give understanding to the events of the patient and they assure that what is happening to the patient does not exclude him from the group. The myths of the group are part of the subjectivity of the patient. It is often the case that the physician is not part of the group and further that his knowledge is seen as threatening whereas the knowledge of the group offers security.

CONCLUSION

For the purpose of understanding I have divided the subjective of the patient into (1) the sociologic person, (2) the unconscious, and (3 and 4) the experiencer and the assigner of understandings. Further, I suggested that understanding is assigned along four planes – (a) timespace, (b) value, (c) causality and (d) association. I think that the physician who is trained to hear in those dimensions will learn how the patient represents his symptoms to himself and assigns them meaning. But further the physician will also learn how the patient will understand future events including what the doctor will tell him of his illness. The patient does not have the option nor an *interest* in seeing things objectively, at least as far as his illness is concerned, because processing the sense data of experience *requires* the subjectivity of consciousness. Details of timing and causality and the adjectives that describe and modify the description of things to come only have meaning to the patient in his own spatio-temporal, causal and value terms. A month may have thirty days, but which is longer, thirty days of health or thirty days of pain caused by my own carelessness?

Two important things remain. First, my division of subjectivity is entirely artificial. Even for the purposes of understanding, these realms and planes should not be seen as parts of a whole. They are not like the muscles, nerves, blood vessels or bones of a limb–things in themselves. Perhaps at their most discrete they are reflecting facets of the whole, a way of beginning to know and appreciate the whole. To see them separately, to listen for them as the patient tells the story of his illness has methodological advantages since the person as subject is so hard to grasp as a whole. The pianist practices the concerto by working out parts of it until those segments are part of him sufficiently so that he can present the whole concerto to himself and then to others. I stress this because two previous attempts to get the subject into medicine represented by the unconscious and the sociologic person foundered because those aspects of person entered medicine as separate and unique entities. Nor have I said what in the person is the assigner of understandings. I do not know and I do not think it is necessary to know. It has been said that a spoken utterance is unique because it can mean literally anything. It achieves its meaning in the moment, the context of other utterances, situation and place and by who says it and who listens. Here also the meaning of events is relative. My divisions and planes of the subjective, as my colleague Dr. Skopek has pointed out, are merely schemes of relationships and in those terms are not totally artificial. But what in the person untangles the relationships is irrelevant.

Finally in this discussion, the interpreter of particulars of the subjective is the physician. Earlier I suggested the image of the physician interposing himself between the experiencer and the assigner of understandings. If he listens only to the sense data of the experiencer he will hear only about disease and will miss the person. If he hears only the subjectivity he will miss the disease. To listen to both without knowledge of disease and knowledge of persons is to miss everything. Have I assigned a role to the physician that I deny to the patient–the ability to keep separate his observation of what the patient says (including how he says it) from his own assignment of understanding or interpretation? Yes, I have done that because I believe that to make *use* of the subjective (rather than just to live it in the everyday) the physician must learn to separate his observation from his interpretation. A difficult task but one that can be taught and learned. We teach the same art in physical diagnosis because we know what we want the student to see and provide a schema with which to evaluate it. Then we hope he will go on to see not only what he has been taught but also, for example, how the skin looks and feels; how the patient walks and how he moves – the endless details of the visible body. The analogy to physical diagnosis is important because there, as in the subjective, we do not expect very much in the beginning. It is our hope, as teachers, that having provided the reason and the tools to see the body, time and the student's ability will enlarge both his knowledge and his views. With the subjective we can only do the same. Provide the reason and the tools to listen and then hope that the physician will hear the person much as the pianist first reads the music and practices individual parts of the concerto. The art of physical diagnosis, then laboratory diagnosis and even clinical thought itself has evolved over the years in a reciprocating relationship to knowledge of disease. The place of the subjective in clinical judgment has rested on intuition for too many centuries. But now it is emerging in response to the demand for a medicine of persons rather than a medicine of disease – a whole medicine, not merely a medicine of the body.

Cornell University Medical College
New York, New York

NOTES

* This work was supported in part by a grant from the Henry L. Blum Research Fund and the Robert Wood Johnson Foundation. With the assistance of Nancy McKenzie.

[1] This essay deals primarily with the subjective of the patient, not the physician. It is concerned with the subjective information that is obtained from patients, not with how that information is processed by the doctor.

BIBLIOGRAPHY

1. Guttentag, O. E.: 1969, 'Medical Humanism: A Redundant Phrase', *The Pharos*, January, 12–15.
2. Harley, J. (ed.): 1967, *Advance Techniques of Hypnosis and Therapy: Selected Papers of M. H. Erickson*, Grune & Stratton, New York.
3. Percy, W.: 1975, *The Message in the Bottle*, Farrar, Straus and Giroux, New York.
4. Weed, L.: 1969, *Medical Records, Medical Education, and Patient Care; The Problem-Oriented Record as a Basic Tool*, Press of Case Western Reserve University, Cleveland.
5. Zborowski, M.: 1969, *People in Pain*, Jossey-Bass Inc., San Francisco.

DANIEL I. WIKLER

SUBJECTIVITY AND THE SCOPE OF CLINICAL JUDGMENT

Subjectivity is a topic of interest to any philosopher, including myself. But *qua* philosopher I am likely to address only a certain group of issues falling under this general heading, relegating others to those of other disciplines. I believe that, for the most part, the questions and claims addressed by Eric Cassell in his 'The Subjective in Clinical Judgment' [1] are not philosophical in nature. Rather, they take the form of a general essay in psychology; and so my comments on Dr. Cassell's discussion of them will be from one in the position of an outsider looking in.

I take Dr. Cassell's thesis to be that subjective reports of patients can be put to good clinical use; that they have not been, so far; and that what is needed is a way of systematizing the subjective in the process of making clinical judgments – as clinicians have already done, in some fashion, with the objective. Further, he holds that patient reports often do not mean what they seem to mean, and that the physician must have a broad understanding of the factors influencing the character of the reports if they are to use them. He makes a number of other points along the way, but I will restrict myself to remarks on these main themes. Stated as they are stated here, I cannot take exception to them, if only because I lack the clinical experience to support any alternative. But such support derives in part from their generality. I propose to examine some of the questions that may arise should the thesis be made more precise and detailed.

The proper place to begin is with the notion of subjectivity – which is, after all, the subject of the paper. One difficulty in reading Cassell's paper, at least in my case, stemmed from his use of certain words which have technical definitions in philosophical writings, used here in some nontechnical sense. The term "subjectivity" – like its opposite, "objectivity" – presents further problems, in that even within the technical literature its meanings are numerous. Some are quite surprising; for example, something possessing what has been called "objective existence" exists solely as an object of thought, fear, or worship – that is, wholly "in" the mind of that thinker. Given the multiplicities of it meanings, I do not think that we can make use of any single standard interpretation of "subjectivity." To avoid misunderstanding, as well as unannounced shifts in content, the term must be defined explicitly.

H. T. Engelhardt, Jr., S. F. Spicker and B. Towers (eds.), Clinical Judgment: A Critical Appraisal, 217–226.

There are at least three meanings of "subjective" which bear on Cassell's discussion:

1. A report may be called "subjective" when it is uncorroborated; when it gives "mere" beliefs and hunches. A casual observer's estimate of the size of a large crowd of people is subjective in this sense. So is a witness's account of the events surrounding a crime, and a patient's recounting of the changes in appearance of a rash he had. This carries the implication that the report is as much a product of the reporter's (patient's) idiosyncratic beliefs and thinking as it is of observations of the facts being reported.

2. Other reports may be called subjective because they report certain facts about the self, especially sensations, thoughts, and other contents of the mind. These are "private events," in philosophical jargon. They may be often detected by persons other than those who experience them, but (so the theory goes) they will be detected by others in a way different from that used by the subject. The patient who becomes aware of an intense pain does so "directly." Others must *infer* the event from other events which they perceive directly – such as a wince, or the patient's report. A patient's report of pain and other symptoms is often subjective in this sense.

3. Though the two senses of "subjective report" mentioned thus far are distinct, they have an important feature in common: such reports are *assertions* or claims, to the effect that something is the case or that an event has occurred. As assertions, they can be true or false, believed or doubted, supported or dubious. In the third of the senses of "subjective" which I mention here, these properties are lacking. A patient's utterance will be "subjective" in this sense when it consists not of assertions but *expressions* of attitude, feeling, or taste. Emotivist moral philosophers, who take ethical statements to be mere expressions of attitude, view ethics as "subjective" in this sense of the term. A "subjective" utterance of this kind will be akin to a frown: though it gives you a way of knowing how the patient feels about something, it need not involve the patient telling you about something.

Which sense of "subjective" is Cassell concerned with? I am not clear on this; perhaps all of these and others as well. In any case, he seems to use the term differently in different parts of his paper. Cassell labels as "clearly subjective" a patient report of an abdominal pain which the patient identifies as a gall bladder disease; perhaps reporting the sensation and identifying it as a symptom of the disease are instances of the second and first sorts of subjectivity which I list, respectively. Elsewhere he explicitly rejects "unreliable" as a synonym of "subjective", using instead (again elsewhere) the paraphrase, "the domain of consciousness of the patient." In one passage he contrasts the subjective with the "objectively confirmable"; in another, with "the objective and measurable."

These allusions and phrasings leave me unclear on which sorts of reports would be counted as subjective by Cassell. When a patient report recounts a memory of a rash, for example, the patient is presumably reporting the rash, not his consciousness of it. If someone else had seen (better yet, photographed) his rash, the report would be confirmable. Is the report then subjective? Again, Cassell devotes a section of his paper to the unconscious as an "aspect of the subjective." Is the unconscious in "the domain of consciousness of the patient" and hence a part of the subjective for that reason? It would seem not, almost by definition; the patient might have unique access to certain kinds of private data useful in learning about his own unconscious (images, urges, etc.), but he will be learning of his unconscious by inference – and thus will be in the same sort of position as the physician. I do not wish to dispute Cassell's claim that the unconscious is an element of the patient's subjectivity; rather, I suggest that the claim cannot be evaluated unless the notion of subjectivity is explicated in some detail.

Indeed, given the variability in meaning of terms like "subjectivity" and "objective", one would expect that without consistent usage, blanket claims about the subjective in clinical judgment may be true relative to some senses of the term while questionable for others. Perhaps one example of this in Cassell's paper is his repeated condemnation of what he takes to be prevailing physician attitudes toward patient reports. Cassell's experience has been that doctors frequently doubt the accuracy of these reports; they find that their patient's understanding is faulty; and they use their own standards, not the patient's, in assessing their value. This observation seems to presuppose what I take to be problematic: that a position either of acceptance or of doubt can be taken toward patient reports as a whole, rather than according to type. If the patient's report is subjective in the first sense I listed – being a hunch or guess, or the product of the patient's idiosyncratic beliefs – it may very well be inaccurate. A patient's subjective reporting of what happened to him during a recent visit to a clinic, or of the history of a present skin rash, must be subjected to doubt by the clinician. It is true that the patient's version of events, however inaccurate, is "true for him." But this is only another way of saying that the patient believes it. False belief on the patient's part may call for a variety of responses from the physician, but trust need not be among them.

We might expect doubt to be unjustified for reports which are "subjective" in the second of the senses listed, viz., reports of private events. That a patient is in pain is, as Cassell states, presumably beyond questioning if sincerity in its reporting can be presumed. But even here, with the reporting of private events, there are distinctions to be made. As Cassell notes, different people use different terms to report what are presumably the same sorts of sensations. This means that the doctor may not get a clear picture of events. Still, doubt is not yet called for. The error lies not in the patient's thinking, but in the communication process. Thus far, Cassell's strictures on doubt are unchallenged. The problem, to my mind, is that there are sources of error, grounds for doubt, which Cassell did not take note of. For example, it is possible that the patient express himself in terms which mean exactly what the physician takes them to mean, and yet, through patient error, not communicate accurately exactly what the sensation is. A term like "intense" behaves semantically like the terms "huge" and "tiny": it makes use of some context-defined reference class. Thus, a huge flea is smaller than a tiny elephant. A patient with limited experience of pain may make a straightforward error in reporting a pain as "intense." More common might be instances in which faulty memory brings a patient to state that he had felt the same pain in the same place before; or, to use one of Cassell's examples, in which a patient claims that an occurrent pain is similar to one which was experienced by another person.

A patient's report which is subjective in the third of the senses I mentioned – being an expression of attitude or feeling – *is*, I believe, immune from doubt. But this is not because it is true or certain. It is neither certain *nor* dubious; for these are properties of assertions and beliefs, not of attitudes or the utterances which express them. The physician may not be pleased with the attitude expressed; he may feel it is unduly negative or unreflective; but there is nothing for him to believe to be *false*. A patient who calls a scar "revolting", or who regards a recommended exercise program as "too much trouble," has not generally asserted anything. The physician can cajole, but cannot correct.

One instance in Cassell's paper in which these distinctions make a difference is in his discussion of the physician's attitude toward a reporting patient's "values." There is an ambiguity in this term between a sense involving attitudes and feelings and a sense involving judgments of quantities or magnitudes which may be of a feeling-neutral sort. Standards of right and wrong conduct are, of course, of the former kind; and it is in the latter sense that one valuates the horsepower of an engine or compares color values of two photographs. A patient who characterizes a pain as "intolerable" may be reporting the perceived intensity of the pain, thus telling us its magnitude, or he may be expressing his attitude toward it (inevitably negative). Were it possible to offer evidence against the patient's appraisal of a pain's magnitude, he would be bound to change his valuation of it. But he would not be bound to change his attitude toward it, for that is not linked by logic to any particular factual claims. Similarly, a person eating oysters may reject them by saying, "I don't like them – they're too salty." He can be told they have no salt, and thus stand corrected (on their "salt value"); but he is still entitled not to like (value) them.

I fear that I have been reading Cassell too literally; for he certainly is aware of the appropriateness of doubt when given a report of symptoms by a patient. Why then, does he employ expressions which seem to argue against it? It seems to me that the source of this lies in the difference between two possible uses of these patient reports. The use which has been the focus of the above discussion has been as a guide to actual events: the physician wants to know what happened (skin rash, pain, and so on), and he hopes to be able to use the patient as an instrument for gaining this information. Used this way, the clinician is interested in such properties as truth and accuracy. The reason for asking for (or listening to) this information is to arrive at a diagnosis and formulate plans for treatment.

In much of the paper, however, Cassell has a different use in mind. He seems to use the patient report, not as a source of information about symptoms, but as a symptom itself, an indicator of certain (usually psychological) properties of the patient which he will make use of in clinical practice. The interest in this case is not in whether what the client believes is actually true, but instead that the client believes it. When patient reports are put to this use, they are again immune from doubt. But this is (also again) not because they are certain – rather, since they are not taken primarily as assertions, adjectives such as "certain", "accurate", and so on cannot apply.

This difference in use of patient reports suggests a parallel difference between Cassell and those he criticizes over the scope of clinical practice and clinical judgment. If one's conception of clinical judgment assigns to it only or mostly the task of forming and confirming hypotheses concerning classification, etiology, and prognosis, one's primary interest will be in forming an accurate record of events; and patient reports will be solicited to that end. This will be insufficient, on the other hand, if one includes within the scope of practice the duty to console and empathize with the patient, and to engineer his compliance to the physician's recommended regimen. That Cassell is in fact interested in these aspects of practice is evident from his writings. This would naturally lead him to an interest in the "true-for-me" as well as in the true. That he finds certain reports valuable which others find only dubitable and misleading may show no more than that he has different interests from the others. And it would follow that what seems at first to be a criticism of the others' evaluative stance toward patient reports may in fact consist of a more fundamental directive to change their view of the physician's role in restoring the patient to health.

The two uses of patient reports are not incompatible; indeed, one may be used as a means to the other. If a patient gives us a symptom report, we may treat it as a datum in itself: as an utterance, with little concern over its truth. If we have an independent means of verifying the report, however, we are in a position to measure the quantity and quality of the patient's coloration of the actual event which constitutes the report's "subjective" aspect. This coloration – report minus the actual record – then becomes the basis for the data we use in determining the patient's psychological traits in which clinicians have an interest. The report, when used this way, is in fact treated as an assertion, just as it would be if we were using it as an instrument for determining what the actual events reported were. But this is done in this instance for the other goal – that is, finding out about the patient rather than about what the patient reported on. Cassell notes, no doubt correctly, that the converse is often valuable; if we come to know enough about the patient's psychological properties (social relations, unconscious material, etc.), we may be able to subtract his idiosyncratic coloration of events from his reports, hence arriving at an accurate record of what is reported. Of course, these remarks hold true only when the patient report is "subjective" in the first or second senses of the term I listed; if it is merely an expression of attitude, nothing is reported; and there is no coloration either to deduce or to subtract.

I think it is accurate to say that what Cassell is offering us in his paper is a program, a proposal-cum-plan for using patient reports which are subjective in some sense of that term. Thus far I've been concerned with interpreting it; at this point I turn to speculation over its chances for success. Cassell's own announced standards are high. He states that the subjective "lies in a domain of medical practice that is least understood, or more precisely that is least systematized," and that "the place of the subjective in clinical judgment has rested on intuition for too many centuries." What he is calling for, then, is a systematic treatment of subjective reporting which will be superior to, and which will displace, the physician's intuition.

Though I am neither clinician nor scientist, I would like to specify why this seems to me to be a tall order. My grounds for this judgment can be introduced through an anecdote. Once, as a child, I was told that I would have an appointment with a physician so that he could treat a minor illness. For some reason I assumed that the doctor would want a careful recounting of symptoms so that he could make a diagnosis, and I took special care to make notes of the various aches and twinges which the illness had brought on, complete with details of their location, duration, quality, and intensity. When I described these to the doctor in as much detail as I could muster, however, I saw that he took no notes. Indeed, my homework was being largely ignored. In retrospect, it is not difficult to discern the reasons for my physician's behavior. It was not that he did not think of me as a person; he was a family friend. It was not that he had not been taught as a medical student to ferret out the subjective in taking histories, for I was handing the information to him without prompting. It is even doubtful that his ignoring my recitation was the result of his doubting the accuracy of my reports. I would imagine, rather, that the reason he took no notes is that he would have had no use for them. I was giving him too much data, too much detail. That my ache occurred intermittently rather than steadily when it appeared the third time told him nothing about my disease. What he lacked was a *theory*, in a broad sense: a set of general propositions about human bodies which would assist him in linking these symptomatic events with medically important bodily conditions. Insufficient medical knowledge stood in the way of making use of my reports, even if they were available and trustworthy.

Just as the physician in this story found no use for the subjective information because he lacked the theory which would enable him to draw conclusions from it, so too might a clinician be disinclined to seek this kind of information from a patient who does not volunteer it. I believe that this consideration is grounds for a certain reserve in the face of Cassell's call for an effort to seek out various sorts of "subjective" information about patients, and I would like to pursue this point with two illustrations drawn from his paper.

The first instance is the use of information about a patient's social milieu as an aid in understanding his symptomatic reports. Cassell states (and of course I take his word for it) that a patient's beliefs about his condition – even his very perception of his symptoms – can be heavily influenced by the beliefs which friends and family members have and have had about their own conditions and symptoms. What does this general truth tell the physician treating his patient as he listens to him telling of his condition and symptoms? In order to make use of the report, he will first have to find out what problems his family and friends have had and what their views of them have been (If *this* in turn is partly a function of what the patient (as family member) thinks about *his* condition, there is a danger of circularity!). The physician must find out what the relationship of the patient is to the group, complete with unresolved unconscious conflicts if possible. Using at this point a general theoretical understanding of the ways in which these beliefs and relationships are related to self-interpretation, the doctor will be able to form a hypothesis about the accuracy of the patient's reports. At the end of this process, he may have acquired some useful information. But does there in fact exist this kind of theoretical understanding? Is there reliable theory on family relationships enabling one to move from data on one quality of the relationship to the pattern of appropriation of others' feelings about their diseases? Without it, there are many missing steps – and the clinician is reduced, after all, to relying on intuition.

The second example, having to do with the patient's perception of temporal distance, follows similar lines. Cassell states, again no doubt rightly, that the perception of passing time, and the location of events in the memory, is idiosyncratic. A patient's reported history will be colored by his own pattern of temporal perception, and only an awareness of this on the part of the physician can enable him to reconstruct the actual sequence of events from the patient reports. In addition, Cassell tells us, the patient is likely to do the same with his own future, and this must be taken into account in management of the patient and his disease. But this general advice, of course, does not tell us how to accomplish this reconstruction (nor does Cassell promise otherwise). For this, we would need a way of determining what a particular patient's pattern of temporal placement is (which must be complex matter); and, in order to project this into the future, we would need to know that this pattern is stable. Without this the reports would be of little use. But Cassell adds that the pattern is *not* stable, after all. It may change in a single patient over time, or even after a single momentous event. Thus, in order to make use of this sort of subjective report, we would need to know, in addition, which momentous events will occur, and we would need a theory telling us generally how patterns of temporal perspective change after such events. Clinicians would be in even a sorrier state if it turned out that the pattern in which the patterns change over time and after important events were also idiosyncratic, requiring further theory in order to make predictions of the patient's perceptions of time.

This is a good deal of theoretical understanding for a clinician to acquire, let alone develop; but without it, we must be left to the unsystematic intuition which Cassell deplores. And the need for theory exists whether the use to which the reports will be put is the traditional one of curing – hypothesis formation, treatment, and prescription – or Cassell's wider goal of "healing" – maneuvering the patient into compliance with the prescribed regimen, assisting the patient in his emotional handling of his illness, and so on. If the goal is the former, the subjective reports will be used as a guide to the discovery of private and public events which might help the physician in his clinical judgment concerning treatment. If it is the latter, the reports will be used as a source of data concerning the patient's psychological make-up and propensities. Both involve hypotheses drawn from clinical observation and processed through theoretical understanding; the latter is evidently just as essential as the former if the goals are to be achieved. A mere injunction to seek out the subjective data will not be productive either of better diagnosis nor of effective handling of the patient unless it can be harnessed for the confirmation of hypotheses of one sort or the other.

What Cassell must require, then, in addition to increased physician interest in subjective reports and data, is theory. It is incongruous, then, to find in his paper a certain reluctance to endorse the utility of this sort of understanding for the tasks which he urges on clinicians. He complains that social data on a patient are used too narrowly, for mere "objective" information: "Sociological parameters do not tell us about *this* divorced woman, but rather about the class of divorced women." This seems to mistake the conclusion of a bit of reasoning for the major premise: from the general truth about the class, plus the information that *this* woman is divorced, comes a probabalistic conclusion about *this* woman: "she is probably. . . ." In any case, the same pattern of clinical judgment must occur even wholly within the "subjective": most Southerners who recount past events this way predict future events that way; ergo *this* Southerner. . . . Theoretical understanding is by nature general; but it seems necessary to use it if the clinician is to make use of available data to learn about his particular patient, whether as subject or as object.

It is also true, however, that the general theoretical propositions will have no clinical utility unless they are used in tandem with the "subjective" data whose collection Cassell urges. Indeed, in the absence of such data, there is little motivation to develop theory that would enable one to use it. Perhaps Cassell's call for increased sensitivity to, and systematization of this information will be the sort of spur which will be productive of its academic development and eventual clinical use.

University of Wisconsin
Madison, Wisconsin

BIBLIOGRAPHY

1. Cassell, E.: 1979, 'The Subjective in Clinical Judgment', this volume, pp. 199–215.

SECTION IV

ROUND TABLE DISCUSSION

JUDGMENT AND METHODS IN CLINICAL JUDGMENT

ROUND TABLE DISCUSSION

STUART F. SPICKER

The term 'clinical judgment' which signaled the principal theme of the symposium which produced these essays and was presumed from the start to be equivocal if not ambiguous has lived up to all expectations. To illustrate this point we need only refer to a few of our previous contributors. Elliott Sober, for example, construes clinical judgment (for his purposes) as ingredient in medical diagnosis, diagnosis having, in his view, no special logic of its own. He asked whether clinical judgment – the result of certain thought processes and/or mental acts–is essentially an art or a scientific enterprise, both or neither. John Gedye understands "judgment", implicit in legal reasoning, as something which can be "mistaken," usually lacks certainty, and is the expression of a "power" we possess which is visibly at work in "diagnostic categorization." Edmond Murphy, as a third illustration, focused on the processes of classification and discrimination. For him, clinical judgment is exercised even in the most basic sortings as, for example, "when the clinician divides his subjects into those with, and those without, coronary disease." Categorization, diagnostic discrimination, discriminant analysis and their concomitant errors all fall, it seems, quite easily under the general rubric "clinical judgment."

To the naive philosopher–one innocent of extensive experience in the clinic and equally unfamiliar with the literature of clinical medicine– the term "clinical judgment" might at the very least conjure the following thought: "Medical personnel perhaps mean by 'judgment' what I have for years taken as expressed by sentences, usually preceded by the term 'that' in typical linguistic form. Hence *It is the case that p, It is not the case that p* are sentences, propositions or judgments which may or may not say anything about the world, any world, this patient or any patient." And here, by way of brief digression, I would appreciate a charitable reading. That is, when I henceforth use the term "judgment" I make no metaphysical assumptions about the existence of such things; thus for me judgment and proposition are equivalent with sentence, the objects with which logic deals. Hence a judgment is true if and only if the sentence that ordinarily would be made by using it is true. Too many advocates of the term "judgment" fail to appreciate a very serious drawback about such entities, as Professor Benson Mates has

H. T. Engelhardt, Jr., S. F. Spicker and B. Towers (eds.), Clinical Judgment: A Critical Appraisal, 229–237.

astutely noted—namely, they do not exist. No metaphysical acuity or mind's eye will ever make them visible. Strictly speaking, then, judgments should not be construed as acts of mind called discourse or reasoning. To permit this is to psychologize logic.

Clinical judgment has a concreteness about it, however. Very generally speaking it encompasses that entire set of judgments or sentences which purport to describe what is or is not the case; in the clinical context this most frequently pertains to some description of a patient's condition. Such judgments include the following: *My patient suffered an attack of acute rheumatic fever.* Again, *This patient presents with aplastic anemia.* Or again, *Prior to final diagnosis it was important to note that a discriminant analysis revealed a murmur of aortic and pulmonary incompetence.* Finally, *The appropriate management of this patient requires a decision to postpone the decision, i.e., we ought to do nothing at this time.*

To avoid remaining naive, the philosopher might have turned to the well-known book, *Clinical Judgment,* which in its earlier formulation appeared in 1964 as a series of separate papers in the *Annals of Internal Medicine* [4]. In this formidable volume Alvan R. Feinstein suggests that clinical judgment is "some sort of intellectual mechanism for organizing and remembering his [the clinician's] observations" ([5], pp. 12-13). In addition, "clinical judgment has a distinctive methodology for dealing with the tangible data of human illness" ([5], p. 28). Furthermore, it is in some sense identical with the observations and reasoning that occur when a physician is engaged in medical problems. Clinical judgment, then, refers to those human methods of evaluating sick people ([5], p. 29). Dr. Feinstein concludes that clinical judgment is a clinical type of reasoning "for making decisions about prognosis and therapy of patients" ([5], p. 12). There may in fact be little reason to challenge these formulations, especially since these expressions of the *sense* "clinical judgment" are shared by other eminent physicians.

When he learned of our symposium, Otto E. Guttentag, M.D., of the University of California School of Medicine at San Francisco (who unfortunately could not attend in person) wrote to me and asked if he might briefly offer this auspicious audience (through me) a few reflections on clinical judgment from the standpoint of the attending physician (in contrast to both the research and the consulting physician). For Dr. Guttentag, I should add, the attending physician is "the key figure of the entire medical effort." He sends us the following message: "From the standpoint of the attending physician, the term 'clinical judgment' refers to the various stages as well as the consummation of the entire diagnostic thought process, i.e., of the attempt to comprehend a patient's actual biological status, (i.e., the most basic and inescapable psychosomatic conditions of his nature) *vis à vis* restoring or maintaining his optimal *biological* status."[1] Furthermore, he adds, "Attending physicians then are not interested in a general *epistemological* analysis of the diagnostic endeavor, e.g., the issue as to whether or not making a diagnosis is an art or a science or both–the less so since the term 'art' embodies a variety of interpretations, e.g., art-state of a discipline, art-craftsmanship, art-capturing the infinite in the concrete. Nor are attending physicians interested in a nosologic-taxonomical analysis of the patient's biological status, i.e., in determining the precise place into which a patient assumedly should be able to fall in a pedigree of diseases, nor in research–quantitating the various sections of such pedigree or of any other reference system. That task belongs to clinical labelers and statisticians, focusing on standardization."[2] Amid all we have heard these past three days, Dr. Guttentag takes the view that "attending physicians are interested in a *phenomenological* analysis of the object of their concern, e.g., the biological and anthropological modes of being in general, their bewildering manifoldness, the subdivision of these modes into being healthy and being sick, and devising a method or methods to manipulate these alternatives." If we take his observation seriously, then a careful account of physician language is in order. That is, we are well advised to enter and eavesdrop in the clinic, no longer, of course, the pre-scientific setting described so well by Michel Foucault in his monumental work on the late eighteenth-century clinic in transition to the contemporary hospital [6]. In the contemporary American hospital what would we overhear? Among and amid the plethora of clinical judgments–no doubt the result of the thought processes and mental acts of physicians–we would easily be able to record a goodly number of value-laden sentences, often unwittingly made under the guise of objective, factual expressions, i.e., indicative sentences which describe the patient's present condition, his past episodes, and the prognosis of states of affairs to be anticipated. Philosophers have already taken pains to alert us to the intrusion of values which infiltrate the allegedly factual language in the clinic. As two philosophers recently put it, "Much would be gained by flushing out these value insinuations from the factual underbrush" ([2], p. 274). There is, then, no need to repeat this admonition here. We need not search for linguistic subtleties when we have yet to attend to the now commonplace judgments uttered in hospital, clinic and even at home (where some patients insist on residing) which are clearly the expression of normative concerns. Quite simply, I call your attention to the language of bioethics or medical ethics which, expressed even more openly in the hospital context, reveals that clinical judgment is now fully infused not only with value-laden notions but quite straightforward moral imperatives, ethical questions, often framed in the language of *ought* and *should* (in their non-prudential usage) rather than by the copulas *is* and *is not*.

When precisely did ethical judgment become so infused and absorbed into the language of clinical discourse? The immediate answer is, of course, "ever since the days of the first physician-patient interaction." Surely physicians always worried over the problem of truth-telling which would almost immediately follow determination of the medical diagnosis. But surely things have changed significantly since the days of Hippocrates of Cos? The modern response is to call attention to the "spectre of technicism" and the technological revolution in which we are presently like flotsam and with which (I fear) we still ineffectually cope.

We could begin by directing our gaze to the invention of the stethoscope and work our way forward to trans-axial computerized tomography (CAT scanner, for short); in between we would find a place for the iron lung (no longer, by the way, an extraordinary life-prolonging treatment) the respirators and ventilators, the renal hemodialysis machines, the shunt for implantation in the myelomeningocele infant's skull, and so on. But such an analysis (in my view) is in fact misguided and does not hit the mark. In fact it serves to camouflage the more salient reason for the rise of moral (ethical) judgments in the clinic and the profusion of ethical sentences into the broader basis of clinical judgments. I call your attention to the powerful decision rule which governs the physician-patient encounter, on the one hand and, on the other, I remind you of the traditional medical ethic (if I may refer to it as such) that until quite recently has prevailed relatively unchallenged.

I deal with the second point first. The traditional medical ethic required that the physician be duty bound to preserve, spare, and prolong human life wherever and whenever he can. He was obliged to fight against disease and death under all circumstances, and was required to do so with whatever was available in the medical armamentarium such as it was ([1], p. 19). The tone of the traditional medical ethic was reminiscent of the tone of the ancient oath of Hippocrates to which the traditional physician swore with binding and fastidious allegiance. It should not go unnoticed that at least three contemporary moral issues in medicine are already foreshadowed in that oath: the prohibition of abortion and euthanasia and the requirement to maintain the patient's privacy and maintain confidentiality outside of the patient's home. Suffice to say that books and articles on moral issues in medicine and health care now proliferate madly. Quite recently Thomas Percival's *Medical Ethics, or, a Code of Institutes and Precepts Adapted to the Professional Conduct of Physicians and Surgeons*, published as early as 1803, has been reprinted, though no longer bound in white buckram [10].

More importantly, perhaps, than the giving way of the traditional medical ethic is the equally ancient, though not systematically appreciated, decision rule that formally governs the physician-patient interaction. Keeping in mind, as Dr. Guttentag reminds us, that the attending physician focuses "upon a modal range of a human being's biologic nature: from optimal to suboptimal status, from being in good health to being in ill health, from being healthy to being ill," ([9]) the physician-patient interaction reveals, when analyzed, essentially four possibilities for the fundamental clinical judgment: to treat or not to treat. The possibilities are as follows (where P = patient):

(1) P is in fact ill and P is judged ill.
(2) P is in fact not ill and P is judged healthy (not ill).
(3) P is in fact ill and P is judged healthy (not ill).
(4) P is in fact not ill and P is judged ill.

Certainly (1) and (2) present no fundamental error of judgment. However, (3) and (4) are errors of judgment on the part of the physician.[3] In (3) the physician dismisses a patient when P is in fact ill; in (4) the physician retains a patient where P is in fact well. The decision rule (or informal norm) which I have in mind, then, is the one devised to handle the *a priori* uncertainty ingredient in the empirical process of medical diagnosis. It is the rule that it is more important to avoid judging a sick person well than to judge a well person sick ([12], pp. 309–313). Those who understand such things will refer to the clinical judgment that *the patient is well when in fact sick* as Type 1 error. The less serious error (but still an error) is to judge that *a well patient is sick.* This is called Type 2 error. This collapses into the decision rule: when in doubt, continue to suspect illness (or disease). To repeat, the error of rejecting the assumption of illness when it is true is far more important to avoid than the error of accepting the assumption of illness when false. From this it is an easy move to the more practical clinical maxim: when in doubt, treat. When in doubt (and that may not occur infrequently) intervene. This practical maxim, grounded at times in epistemological vagueness and accompanied by severe professional sanctions at others, serves to reinforce the traditional medical ethic. A physician who fails to treat a patient who subsequently dies from the undetected condition is not only subject to legal reprisal (or should I say legal *action* for negligence) but may be open to moral sanction by peers and public. Now add to this state of affairs what Victor R.

Fuchs calls the "technological imperative," namely, "the desire of the physician to do everything that he has been trained to do, regardless of the benefit-cost ratio" ([7], p. 60). In short, "If you have it available, then use it; if it exists in the armamentarium of medicine, employ it." In addition, stir in Lewis Thomas' putative claim that "If we have learned anything at all in this century, it is that all new technologies will be put to use, sooner or later, for better or worse, as it is in our nature to do" ([13], p. 79). Now combine all ingredients with the decision rule – "when in doubt, treat" – and is it any wonder that public sentiment is the resultant of forces which currently combine to compel us to assess the traditional medical ethic, since it is surely in conflict with the even more ancient physician mandate – *primum non nocere*, first, do no harm. The contemporary worry, then, is that in far too many cases medicine, taken in the broad sense, in pursuing an imperative to treat is now the perpetrator not only of injury but also of harm and even wrong.

Clinical judgment, when articulated as a particular judgment – *I am in doubt about the outcome of my examination, therefore I suspect illness*, has resulted in the tendency to treat and even overtreat. This, accompanied by the physician's ability to bring the most ingenious machinery to bear on various forms of human illness, generates the entire spectrum of bioethics.

To gain a more wholesome perspective we will have to begin to resist the traditional medical ethic and argue well for its inappropriateness in particular cases. We will have to temper, even more splendidly, the gauge of clinical discrimination, since it now universally includes normative considerations which reinforce the fact that medical or clinical judgment is essentially a normative, i.e., ethical, enterprise which of necessity includes quite unambiguous moral judgments. We can no longer rest comfortably with surreptitious and informally unofficial clinical judgments which are, as I suggested, as moral as they are medical. For example, it is not uncommon any longer for the intern, under the auspices and supervision of the attending physician, to write 'NO CORE' or 'DNR' (do not resuscitate) on the order sheet of the patient, a notation which eventually finds its way to the kardex, the final locus of the up-to-date orders of the attending physician as transcribed and signed by a secretary, clerk or registered nurse. The notation 'DNR' clearly includes, albeit tacitly, the judgment that *we should desist from further treatment of the patient* (where the letter 'R' for 'resuscitate' is employed in the broad sense so as to include all life-prolonging interventions and not just the discontinuance of the resuscitator). Such notations, which at present have no formal sanction or official legal status and (as far as I know) are untested in the courts, forestall the important dialogue concerning the ethical issues ingredient in a decision to withhold treatment. It may be that refraining from further treatment and clinical intervention (as many now tell us) is the morally appropriate course of action in many cases, but whether or not it is, it is in need of exploration. That exploration may require the participation of persons formally trained in ethical analysis. That person might well be the physician in time to come.

In closing let me say that medical ethics has for some time ignored nursing ethics, the first formal text being published in 1903.[4] Amid the numerous definitions of the normative science of ethics which proliferate the philosophical and medical literature, one offered by Isabel Hampton Robb, an outstanding nurse who practiced at the turn of the century, is poignant for our purposes. I cite her definition: Ethics is ". . . the science that treats of human actions from a standpoint of right and wrong. It teaches men the practice of duties of human life and the reasons for what they do *and for what they should leave undone*" ([11], [3], p. 174). I stress the last clause, for it is equally important in the domain of conduct to know when to do nothing and what to leave undone. If this judgment is as important to medical practice in the contemporary period as I believe it is, and if, as Dr. Edmond Murphy suggests, practitioners "must make decisions even if the decision is to postpone the decision or to do nothing," we may find that we are on the way to a new medical ethic, or at least the traditional medical ethic is in the process of undergoing radical transformation. The most frequently discussed moral issue in medicine is the concern that many have that it may now be morally obligatory, in many cases, to desist and refrain from treatment and, if you like, to withhold it. At times this includes withdrawing it. Today not only the traditional medical ethic is undergoing metamorphosis but so too is the decision rule which continues to govern the very foundation of clinical judgment, that is, the very *Ursprung* of reason in medicine.

University of Connecticut Health Center
Farmington, Connecticut

NOTES

[1] A personal communication, dated March, 1977, 2 pages.

[2] *Ibid.* Also see [8].

[3] I leave aside here such special cases as the patient who in fact presents ill, but the physician's judgment is not to administer any drug regimen. This, for me, is still a case of treatment, though the treatment is to *refrain* from prescribing all medication. Hence this case does *not* entail physician error and easily falls under permutation (1). Again, I am

untroubled by the case (suggested to me by Samuel Rabison, M.D., to whom I am indebted for a careful reading of this paper) of the child who (accompanied by his parents) presents with cough and mild upper respiratory distress and is given antitussin *pro re nata*, though this medication is prescribed to assuage the worry and anxiety of the parents, with no real medical benefit in itself. This is simply a case under permutation (1) i.e., the child is indeed mildly ill and the physician elects to treat by means of a placebo of sorts, since he or she adjudges that no medication is warranted, e.g., antibiotic therapy, and chooses therefore, as treatment regimen, to allow natural processes to occur which, it is believed, will result in improvement of the child's condition. The case for construing permutation (4), Type 2 error, as no error at all, may be the one in which the healthy child is simply taken to the doctor for a check-up or for reasons of securing a "bill of health" for school purposes. Here *ex hypothesi*, the child is not ill, and yet the physician may recommend some substance or dietary regimen because of the social pressure to "do something". But this is not a counter case under (4) but simply one which easily falls under (2). For the physician in this case is not, strictly speaking, providing medical treatment, but satisfying parental need (regrettably too common a practice).

[4] Eleanor Crowder cites this definition. See ([3], p. 174). Also see [11].

BIBLIOGRAPHY

1. Berg, J. H. van den: 1969, *Medische Macht En Medische Ethiek*, G. F. Callenbach N. V., Nijkerk, London.
2. Clouser, K. D. and Zucker, A.: 1974, 'Medicine as Art: An Initial Exploration', *Texas Reports on Biology and Medicine* **32**, 267–274.
3. Crowder, E.: 1974, 'Manners, Morals, and Nurses: An Historical Overview of Nursing Ethics', *Texas Reports in Biology and Medicine* **32**, 173–180.
4. Feinstein, A. R.: 1964, 'Scientific Methodology in Clinical Medicine', *Annals of Internal Medicine* **61** (Nos. 1–4), 564–579, 757–781, 944–965, and 1162–1193.
5. Feinstein, A. R.: 1967, *Clinical Judgment*, Robert E. Krieger Publishing Co., Huntington, New York.
6. Foucault, M.: 1973, *The Birth of the Clinic: An Archaeology of Medical Perception*, transl. A. M. Sheridan Smith, Random House, New York.
7. Fuchs, V. R.: 1974, *Who Shall Live?: Health, Economics, and Social Choice*, Basic Books, New York.
8. Guttentag, O. E.: 1949, 'On the Clinical Entity', *Annals of Internal Medicine* **31**, 484–496.
9. Guttentag, O. E.: 1978, 'Care of the Healthy and the Sick from the Attending Physician's Perspective: Envisioned and Actual (1977)', in S. F. Spicker (ed.), *Organism, Medicine and Metaphysics: Essays in Honor of Hans Jonas on His 75th Birthday*, in the Series, *Philosophy and Medicine*, VII, D. Reidel Publishing Co., Dordrecht, Holland/Boston, Mass., pp. 41–56.
10. Percival, T.: 1975, *Medical Ethics*, Robert E. Krieger Publishing Co., New York.
11. Robb, I. H.: 1903, *Nursing Ethics: For Hospital and Private Use*, J.B. Savage, Cleveland, Ohio.
12. Scheff, T. J.: 1971, 'Decision Rules and Types of Error, and Their Consequences in

Medical Diagnosis', in Eliot Friedson and Judith Lorber (eds.), *Medical Men and Their Work: A Sociological Reader*, Aldine-Atherton Press, Chicago, Illinois, pp. 309–323.

13. Thomas, L.: 1974, *The Lives of a Cell*, Bantam Books, New York.

ROUND TABLE DISCUSSION

E. JAMES POTCHEN, PAUL WAHBY,
WILLIAM R. SCHONBEIN, AND LINDA L. GARD

We assume that clinical judgment is a complex, multi-dimensional process. We will attempt to examine the process in terms of one dimension of the patient-physician interaction, i.e., that dimension which relates to the generation and use of clinical data. We propose that although the flow of data between the clinician and his patient is meaningful, the process by which data becomes information is not necessarily clear. It will be our task, then, to develop a model for assessing data and information in an attempt to unpack and expand upon the phenomena of sound clinical judgment.

In proposing such a model, we will assume that the physician-patient relationship is purposeful and that this purpose, in part, relates to the physician's attempt to gather data about a patient in such a way as to allow for an improved capacity to predict the effects of intervention in the course of a patient's disease. The probability that a specific intervention will alter the course of the disease is dependent upon what is wrong with the patient. In this context, then, we can regard clinical judgment as the capacity to utilize data. We use the term judgment to mean a mental ability to perceive and distinguish relationships and alternatives. We stipulate that reasonable judgment entails the generation and use of data in such a way as to beneficially affect the outcome of what happens to a patient. Through mental process which we will subsequently seek to define, purposeful data is chosen from a multitude of possible attributes manifested by the patient. The selection of data elements (patient attributes) relevant to clinical decision-making is a basis for considering the judgment of a physician.

In an effort to develop a model of clinical judgment in diagnostic decision-making we will first have to model the process of medical diagnosis. Here we use the term "model" to mean a set of idealized assumptions about the inner structure, composition, or mechanism of a system. A patient can be looked upon as an open system where numerous possible data elements can be presented patient attributes. Figure 1 illustrates the flow of data from the patient to the physician and its ultimate relationship to action and outcome. Here we see that some of these attributes are initially apparent to the physician in terms of the patient's complaints. Others are sought after in the history, physical, and selection of "appropriate" laboratory tests in an effort to clarify

H. T. Engelhardt, Jr., S. F. Spicker and B. Towers (eds.), Clinical Judgment: A Critical Appraisal, 238–247.

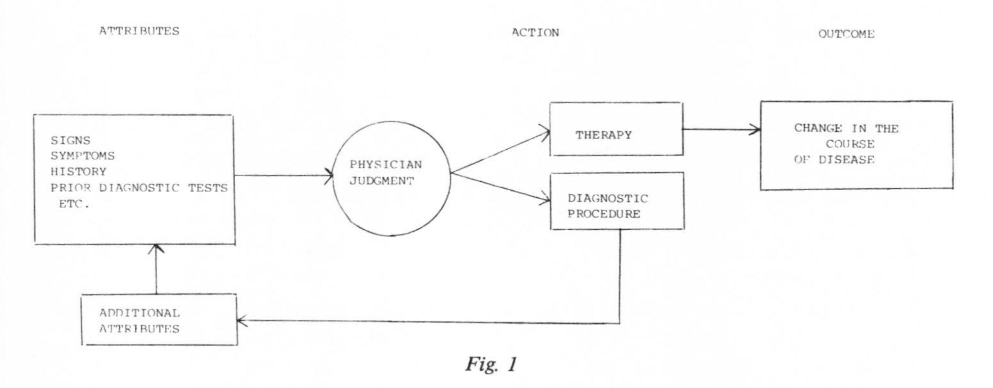

Fig. 1

the issues sufficiently enough for the physician to effect a judgmental decision as to whether additional data is warranted, or if the necessary data has been provided to direct an intervention in the course of the patient's disease. In the aggregate of patient attributes, the physician discovers a pattern which influences his judgment and ultimately his behavior. The amount of confidence placed in each element of data about a patient is dependent in part upon other attributes impinging on the system. Very few single patient attributes (complaints, signs, symptoms, or diagnostic tests) are specific for a single disease process, much less the eventual outcome of the disease. Rather, most of the individual data elements collected by physicians are associated with many possible outcomes and the weight assigned to each data element may be influenced by other features of the judgmental process, e.g., decisions regarding value, thinking processes, etc.

The value of individual diagnostic tests in medicine has largely been studied in relation to the probability of their occurrence in a specific disease process. In terms of sensitivity, diagnostic test A is considered more valuable than diagnostic test B in detecting disease D where test A occurs with greater frequency than test B in disease D. The term "sensitivity of a diagnostic test" refers to the frequency with which it occurs in a specific disease. "Specificity" of a diagnostic test refers to its failure to occur in diseases other than the postulated disease process. The sensitivity and specificity of diagnostic tests in medicine have most often been studied by relating a *single* individual test (data element) to their presence in a specific disease. While this approach has been useful in defining the relative merit of alternative diagnostic strategies, it does not replicate the process observed in medical decision-making. Judgmental decisions in medical diagnosis are rarely limited to utilization of a single piece of evidence. Rather, the physician has the opportunity to consider multiple data elements simultaneously and in series. Through a thinking process he uses discretionary judgment to select and/or exclude various data bits from his decisions. Judgment, in part, depends upon a capacity to filter data. The weight given any single data element is, in part, dependent upon the presence or absence of other data elements. Effectively, the process is not one which can be adequately understood through a single single-variate analysis. Rather, judgment in diagnostic medicine is a multi-variate process requiring analytic tools and models appropriate to its multi-dimensionality.

Elstein has shown that "rather than follow a set order of constraining questions, the experienced physician appears to leap directly to a small array of provisional hypotheses very early in his encounter with the patient" ([5], p. 86). This observation suggests that the process of judgment is essentially an iterative technique whereby specific attributes present clues which define the hypothesis that the physician would subsequently test by seeking confirming attributes. These defining and confirming attributes are replicated *ad seriatim* in a unidirectional, ongoing and dynamic process that perpetuates itself toward an increased state of knowledge about the patient and the disease. It is apparent, though, that no single clue is necessary and sufficient for the physician to predict an outcome or direct an intervention. Instead, he identifies clusters of clues which generate his hypotheses. For example, the presence of a sore throat, per se, is insufficient for the physician to warrant the conclusion of a streptococcal infection whereby the patient will benefit from the administration of antibiotics. Rather, at the time he hears of the patient's sore throat, the physician seeks additional data concerning temperature, and even confirmatory diagnostic tests (such as bacterial culture) before he concludes that he has identified the probability that an intervention, in this instance the administration of antibiotics, will significantly benefit the patient. In this case, there are multiple attributes: the patient's complaint of a sore throat, the physician's observation of a throat which is consistent with streptococcal infection, the presence of a fever, the duration of the illness, and even the more definitive, confirming throat culture. While obtaining these apparently relevant data elements, the physician has also obtained numerous other attributes about the patient such as age, sex, and the presence or absence of lymph nodes in the neck. One can appreciate that this rather simplistic example is often considerably more complex when the physician is forced to make a judgment concerning the merits of additional diagnostic tests in addressing a patient's problem.

Thus, the experienced physician appears capable of identifying the cluster of attributes most meaningful in verifying his hypothesis and selecting from the broad diagnostic armamentarium those techniques which are most useful in enhancing the probability that an intervention will benefit the patient. It seems, then, that judgment in diagnostic medicine entails the ability to perceive and distinguish relationships or alternatives when faced with a mass of potentially meaningful data elements, and placing a value on specific, future data elements, i.e., additional diagnostic tests, based upon the discovery of a pattern or cluster in the attributes presented by the patient. It is the physician's task to discern which data elements are meaningful when considered *in toto*. Therefore, in developing a model of clinical judgment in medical diagnosis, one must be able to look at multiple variables simultaneously and ascribe relative values to their inclusion or exclusion as meaningful components affecting a diagnostic process. The detection of diagnostic clues from the available data provided to a physician requires a thought process which is multi-variate; that is, it deals with more than one variable and simultaneously weights them according to the presence or absence of numerous other variables. For this reason, we choose to differentiate between mechanistic models which are distinguished in their coupling or cascading of forces appropriate to the level of organization in which the model is used versus entropic models which have been related to the measurement of a quantity called information.

The word "entropy" was derived by Clausius in 1854 from the Greek "a transformation", and has been called the transformation content of a body or system whereby every irreversible change in the system will increase its entropy ([3], p. 116). Though initially formulated as a mechanistic interpretation of such physical phenomena as heat exchange and the movement of ideal gases, the inherent paradoxes of such an interpretation soon became evident. For instance, it was argued that if matter is finite and time is infinite, then there exists a finite number of configurations of matter since there is essentially "time" enough for all possible combinations to arise and eventually reoccur. In the event of a cyclical reality, then, the net gain in entropy must be zero. Thus, the concept of entropy was developed by Boltzmann and Planck to assume a more statistical nature based on probability, not causal certainties ([3], p. 119). In this way, the universe could have a net movement toward disorder, and yet at the same time allow for low-probability, entropic anomalies such as life in our solar system. The statistical basis of entropy provided the groundwork for Maxwell's introduction of the "sorting demon" in 1871 ([3], p. 162), and ultimately led Leo Szilard to formulate the concept of "information" of entropy ([3], p. 161).

Modern information theory is derived from the work of Claude Shannon in 1948 whereby information is concerned with the statistical character of the whole range of possible messages ([3], p. 161). While entropy originally signified the measure of the amount of disorder in a physical system, it can be appropriate to measure the direction of change in knowledge about the structure of any system. The difference in the entropy of a stochastic system and that of an ordered counterpart is a measure of the information extracted from the system as a result of imposing some order upon it. Thus an increase in "negative-entropy" corresponds to an extraction of information in that it represents some decrease in the randomness of the data and signifies an increase in the state of knowledge of the system.

With the development of information theory and an awareness that the statistical concept of entropy originating from thermodynamics can be associated with complex multivariate systems, an opportunity was provided to apply these principles to the data gathered from any complex system. Patients present with multiple elements of data, some of which provide information relevant to an outcome of their disease process. The judgment of a physician is applied toward seeking an increase in negative-entropy from components of the data when faced with the entire array of possible patient attributes. In effect, it is feasible to look on this mathematically as an N-dimensional matrix where totally random data elements contribute no information to predicting the probability of an outcome state, whereas clusters of less than total randomness are increasingly informative in defining a disease process for the physician. One could then look upon clinical judgment in medical diagnosis as a process of sorting and analyzing available data in order to formulate a hypothesis concerning the probability of the physician's behavior affecting the patient. The value given to a clue by a physician depends in part upon how randomly the clue, which represents an aggregate of data elements, defines a hypothesis and warrants subjecting the patient to the elicitation of confirming attributes, i.e., further diagnostic tests, or prescribing an intervention which will alter the course of the patient's disease. If all the data elements were merely randomly correlated with outcome, they would have relatively little meaning in being able to anticipate whether or not such a test or intervention would benefit the patient. It therefore develops that a model of judgment in medical diagnosis would relate to the capacity of the physician to select the aggregate data elements in the formulation of clues which predict an outcome state. A simple test of this model would be to look at the number of data elements provided by patients in a specific disease process and test the physician's judgment concerning implications of these clues in predicting an outcome.

Bell and Loop in 1971 observed the phenomena of skull X-ray requisitions among a community of physicians in Seattle ([2], pp. 236–9). They developed an encounter form which listed the data elements provided to the physician in a course of caring for patients with head injury in an emergency room. They asked the physicians to predict the probability of finding a skull fracture on the basis of their analysis of the available elements of data. This data has been subjected to an entropy mini-max multivariate analysis technique in an effort to identify the relative value in terms of predictive significance of each data element when the entire aggregate of data elements is observed simultaneously. This technique entails a computerized monitoring for data clusters by imposing "partitions" in N-dimensional space that correspond to specific disease outcomes. Since random data should produce a haphazard array in N-dimensional space, data that correlates with a specific outcome will occur in some pattern or cluster, thereby denoting some lack of randomness. It is this deviation from the maximal entropy of the data that

TABLE I

Attributes Affecting Physician's Subjective Probability of Skull Fracture
Attributes having significant effect (approximate order of importance)
1. Evaluation of injury
2. Drowziness
3. Conscious/unconscious
4. Consciousness scale
5. Breathing
6. Sensory Abnormality
7. Ear discharge
8. Eardrum discoloration
9. Nose discharge
10. Headache
11. Palpable bony malalignment
12. Swelling
13. Hematoma
14. Retrograde amnesia
Attributes not appearing in cell definitions
1. Vomiting
2. Seizure
3. Other reflex abnormalities
4. Anisocoria

signifies a net gain in information. The "entropy minimax" method attempts to *mini*mize the entropy of a partitioning thereby *maxi*mizing information extraction from the data.

Table I lists the attributes which affect the physician's subjective probability of a skull fracture in this population of patients. These attributes are in the approximate order of importance in affecting his judgment as to whether or not a fracture would be present prior to the defining attribute, i.e., skull X-rays, which confirmed or failed to confirm the physician's judgment. The entropy mini-max approach allows for hierarchical listing of the relative value given each piece of data by the physician. The use of this data as informative for outcome is given a subjective weight by the physician when observing the other data in context. The technique also allows for the definitions of attributes which do not appear in cell definitions, i.e., attributes which are not used by the physician in detecting the cluster of relevant data elements affecting his judgment.

TABLE II

Correlation of Attributes With X-Ray Outcomes of Skull Fractures
Attributes appearing in cell definitions (approximate order of importance)
1. Evaluation of injury
2. Consciousness scale
3. Conscious/unconscious
4. Sensory abnormality
5. Ear discharge
6. Other reflex abnormality
7. Palpable bony malalignment
8. Headache
9. Nose discharge
10. Vomiting
11. Anisocoria
12. Hematoma
13. Drowziness
14. Retrograde amnesia
15. Swelling
Attributes not associated with cell definitions
1. Seizure
2. Eardrum discoloration
3. Breathing

Table II permits a comparison of the physician's subjective probabilities with the correlation of attributes which indeed affected the X-ray outcome. In this instance, the data was subjected to an entropy mini-max analysis where the outcome state was the presence or absence of a real skull fracture seen on the X-ray. By comparing the hierarchical list of attributes that the physician uses in making his clinical judgment with the second list of attributes that correlates with the presence of a fracture on the skull X-ray, we are able to discover the discrepancies between clinical judgment and diagnostic reality. The entropic model of clinical judgment thereby provides for an opportunity to evaluate qualitatively the judgmental inference physicians derive from the available data elements supplied by the patient in relationship to a reality test defined in terms of the cluster of elements which do in fact influence the outcome. It is interesting to note in comparing these two hierarchical lists that physicians put a relatively high value on breathing pattern which is not correlated as valuable in defining the presence or absence of fracture. On the other hand, physicians ignore vomiting, which is an element that does affect whether or not a skull fracture would be found.

The entropic model of clinical judgment allows for better understanding of the distinctions between the judgment of a physician and diagnostic reality, but has significant limitations when applied to the many features of the complex problem of clinical judgment. Purely rational judgmental process would demonstrate a similarity between what the physician's subjective probabilities are versus the diagnostic realities when correlated with a specific outcome. This difference between judgment and reality measures is in part related to other factors that affect the clinical judgment of physicians in addition to the clinical data which comes from the patient. Table III defines a model of clinical judgment which depicts the distinction between patient-dependent and physician-dependent elements of the clinical judgmental process. It is important to point out that medical decision making which appears as clinical judgment is not limited to decisions on clinical data. Rather judgmental decisions relate to the patient's "true" value system and the physician's perception of the patient's value system in addition to the physician's perception of clinical data. Finally, clinical judgment is influenced by the physician's value system and his/her thought process. The unpacking of thinking processes seen in Table III is modeled after Koestler ([7], pp. 630–1).

A purely rational system would be limited to a definition of attributes, utilization of appropriate attributes to lead toward intervention, and a prediction based on probability concerning a desirable outcome state. However, it is apparent that judgment is significantly influenced by many other factors

TABLE III

Features of Clinical Judgment
Patient dependent features
1. *Clinical data* signs, symptoms, laboratory, x-ray, etc.
2. *Patient's value system* religious, economic, social, health considerations, etc.
Physician dependent features
1. *Extrinsic* a. Physician's perception of clinical data and the patient's value system
2. *Intrinsic* a. Physician's thinking processes affecting his/her behavior 1) degrees of consciousness 2) degrees of verbalization 3) degrees of abstraction 4) degrees of flexibility 5) type and intensity of motivation 6) realistic vs. autistic thought 7) dominance of outer and inner environment 8) learning and performing 9) routine and originality

held within the physician at the time of the physician-patient interaction. These features can significantly affect the physician's response to data, his capability of receiving and distinguishing the relationships between data and information, and his decisions concerning the gathering of further data or intervention in the disease process.

The attempt to understand the process of clinical judgment must begin with the identification of its essential components. We have seen that although correct medical diagnosis is bound up in the multi-dimensional analysis of data presented as patient attributes, it can be strongly influenced by the subjective interpretation of the physician. Through a process such as the entropy mini-max multivariate analysis, we can formulate a stoic rationality in clinical judgment. This technique will provide the opportunity for greater insight into the factors which affect judgment outside of a rational information process and may lead to an enhanced awareness of numerous features of clinical judgment which are not dependent upon the data emanating from the patient. Thus, we regard the entropic model of data and information to be a powerful tool in the task of understanding the process of diagnostic decision-making and clinical judgment.

Michigan State University
East Lansing, Michigan

BIBLIOGRAPHY

1. Bazarov, I. D.: 1964, *Thermodynamics*, trans. by F. Immirzi, Oxford University Press.
2. Bell, R. and Loop, J.: 1971, 'The Utility and Futility of Radiographic Skull Examination for Trauma', *New England Journal of Medicine* **284**, 236–9.
3. Brillouin, L.: 1956, *Science and Information Theory*, Academic Press Inc., New York.
4. Cohen, I. B.: 1974, in F. Suppe (ed.), *The Structure of Scientific Theories*, University of Illinois Press, Urbana, Ill., p. 308.
5. Elstein, A. A., et al: 1972, 'Methods and Theory in the Study of Medical Inquiry', *Journal of Medical Education* **47**, 85–92.
6. Jones, D.: 1977, 'Entropic Models in Biology – The Next Scientific Revolution?', *Perspectives in Biology and Medicine* **20**, 285–98.
7. Koestler, A.: 1967, *The Art of Creation*, MacMillan Company, New York, pp. 630–31.
8. McNeil, B., et al: 1976, 'Primer on Certain Elements of Medical Decision-Making', *New England Journal of Medicine* **293**, 211–21.
9. Schonbein, W. R.: 1975, 'Analysis of Decisions and Information in Patient Management', in E. J. Potchen (ed.), *Current Concepts in Radiology*, Vol. 2, C. V. Mosby Company, St. Louis, pp. 31–58.
10. Shannon, C. E. and Weaver, W.: 1962, *The Mathematical Theory of Communication*, University of Illinois Press, Urbana, Ill.

SALLY GADOW

Beyond the epistemological problems of whether clinical judgment can be scientific, subjective, and logical, there is the profoundly human problem: How, if at all, is clinical judgment to be integrated into the developing life-forms of the individual? What has clinical judgment to do with the individual's life? The answer is often "very little", precisely as medical thought becomes more sophisticated and objective, even as it indulges in pre-systematized, enlightened subjectivity. The problem known to anyone who has been either patient or clinician, is that the carefully wrought and professionally administered medical judgment is all too often received by the patient as being either incomprehensible with respect to its personal significance, impossible to believe, the absolute truth, or – perhaps worst of all – irrelevant. In all of these cases the patient leaves the implications of the conditon where they seem to have originated: with the clinician, either by submitting in helpless defiance or trustful deference, or by literally leaving (against medical advice). In none of these cases is a relation established between the person and the condition which sustains the human individuality involved.

In the following discussion, I hope to show how this tendency evolves, and to propose an alternative, more philosophically complete meaning of clinical judgment for the individual, based upon the dialectical method described by Hegel. I will suggest that the suppression of the individual qua individual by medical judgment will not be remedied by making the clinician either more scientific or more humane, more infallible or more tactful. Nor will the individual be affirmed by making clinical judgment, in this case self-diagnosis, the solitary task of the patient.

Before beginning, let me briefly recall the sense of 'philosophically complete' which Hegel articulated and which I will use here. For Hegel the complete understanding or judgment of a situation is one which achieves its completeness through a process of development. And this process, to reach true completion, must be dialectical: each successive stage comes about as an overcoming of the specific defect in the former stage. That overcoming can proceed in different ways. At early stages of a dialectic, brute opposition occurs, one defect succeeded by its opposite, its mirror image, so to speak. As the dialectic advances, its negations are more positive, so that a defect or

H. T. Engelhardt, Jr., S. F. Spicker and B. Towers (eds.), Clinical Judgment: A Critical Appraisal, 248–253.

one-sidedness is overcome by incorporating rather than excluding the exaggerated element. The essential point in Hegel's concept is that it is not the outcome but the total process of dialectical development which is philosophically complete. The conclusion itself has neither truth nor value except as having evolved out of the dialectic.

It is thus the process of development of the clinical judgment reached in each case which will constitute a philosophical judgment, complete in Hegel's sense if its development is a dialectic that overcomes, not omits, opposing elements. Conversely, when any element is not resolved but simply negated by exclusion, the process remains incomplete, the judgment partial, and medical thought loses itself in extremism. To elaborate the nature of the judgment that I propose as philosophically complete is at the same time to describe its dialectical evolution–and thus to describe a series of stages and transitions, each one of which can become a premature endpoint. In describing the process, then, which is necessary as a totality, I will be relating the distinct levels and possible dead-ends to which I referred when I mentioned the patient (and I now include the clinician) who regards a clinical judgment with either skepticism, deference, or personal indifference.

The earliest level of the dialectic, in the sense of most immediate, is the experience and judgment of the symptom. The individual expresses feelings, or the clinician observes phenomena, as directly constituting rather than signifying a disorder. Whatever the symptom–pain, cough, visual loss–*it itself* is the disorder; it does not signify a condition beyond itself. Its immediate, non-referring character is expressed by the person who demands alleviation of distress without any interest in underlying processes or causal entities. Such insistance is most vehement, of course, in the protests of the sick child, who wants nothing of theories, thumping, and laboratory procedures. The same immediacy is expressed by the clinician in the simple descriptive statement of the symptom ("You have a fever") and the course of action directed toward the symptom as the final object of treatment.

The problem with this level is obvious: symptoms are not always alleviated nor patients restored by "symptomatic" treatment. In seeking to remedy this, medical thought leaves the immediacy of the experiential, the symptom per se, and judges the same phenomenon as being *sign*, referring not to itself but to a concealed meaning, undisclosed until the sign is deciphered. Experience becomes significant for a new reason: not only because it is distress, but because it signifies. "Your fever indicates infection." Fever is not a problem in itself to be immediately relieved because of the discomfort it brings; it is an important cipher which must be allowed to proceed unimpeded in order to disclose its special meaning, for different febrile patterns conceal different disease realities.

But the patient's experience, though indicative of disease, also interferes with access to the disease. Perfect comprehension of the disorder requires abstraction from all of the empirical variations confounding the purity of the meaning, the disease essence. "The individual must be *subtracted* to understand the disease" ([1], p. 14). At this point in the dialect, when a metaphysics of nosology prevails, both patient and clinician as individuals are sources of distortion in the expression, perception, and judgment of the disease in its pure form. Age, temperament, error, and all other idiosyncracies must be negated on both sides, for the truth of medical judgment is now its signification of universals. At this stage, clinical thought strives to overcome completely the immediacy of the phenomenon as symptom, the judgment of which could be verified only by the patient's experience. Now, the clinical gaze strains for the thing-in-itself which is disclosed only obliquely in the empirical body, and may be altogether inaccessible to the patient who cannot decipher the signs or look beyond the urgency of the body's experience.

What this level of thought does not overcome, however, is the immediacy of the opposition between essence and accident, diseases and individuals. In negating the non-signifying immediacy of particular symptoms, medicine generates an untenable contradiction between its nosological types and the particular, ever-varying incarnation of those types. It is not only ineffective to treat physical phenomena as deviations from a metaphysical reality. It is contradictory to treat with material means an ideal condition. The pure case, like Plato's ideal triangle, never appears, not because perfect participation of forms in phenomena is infinitesimally improbable, but because it is a priori impossible.

Neither medical theory nor practice can proceed on the basis of such a fundamental contradiction. Essences are thus overturned by a new level of thought, a clinical empiricism in which the individual is no longer an instance of a disease species "inserting itself into the body. . . It is the body itself that has become ill ([1], p. 136). There are no transcendent entities; there are only visible processes. And those processes, though classifiable, are infinitely varied in their appearance, and can be described without reference to personal context or disease species. Hence the authoritative judgment is that of neither the patient nor the clinician, but the pathoanatomist: the particular lesion itself, different in every case, is the new and ultimate ground of medical certainty.

But the overthrow of essences is in effect followed only by the opposite form of transcendentalism, for the lesion can be silent, inaccessible to personal awareness or clinical observation. If not silent, it may in its uniqueness defy inference on the basis of the signs it manifests, awaiting autopsy examination for its actual form to be revealed and the conclusive judgment expressed. The pathological particular, like the universal, can be abstracted so far from the individual human life that it transcends the possibility of being experienced. At neither extreme, then—medical essentialism or empiricism—does the individual participate in or otherwise directly relate to the process of clinical judgment.

The form of medical thought which attempts to unite these extremes and focus on the complex totality of the individual person and situation is the level I will call probabilistic medicine. Here judgment is not based upon abstract essences. In the behaviorist idiom, there is no such entity as diabetes; there are only persons with insulin deficiency. Similarly, judgment is not based upon the abstract particular, the lesion. There are persons with lesions, but persons who also have families, anxieties, histories, and values. All of the particulars in an individual case have equal value initially. None is discounted; each is capable of becoming associated with a recurring configuration that is then applied in describing, identifying, and predicting the condition. The judgment of 'the condition', therefore, is as comprehensive and as individualized as the quantity of data permits, and includes particulars from as many realms of human existence as can be scanned in a given case—anatomic, biochemical, genetic, developmental, economic, psychosocial.

At last the individual begins to emerge as the object of clinical judgment—but the individual as assembled by the data analyst, not as encountered by the clinician. "Medical certainty is based not on the *completely observed individuality* but on the *completely scanned multiplicity of individual facts*" ([1], p. 101). The individual is not seen as a self-unifying whole but a multiplicity of events thematized by frequencies. And that task, of external unification on behalf of the individual, can never produce a philosophically complete judgment because the array of particulars is infinite and, seen from the outside, is without any order until one is provided for it. The premise of probabilistic medicine, that individual variables are more important than prior clinical generalization, is a crucial step in the dialectic, but it itself is a generalization which assumes for medicine the responsibility for deciding which variables and idiosyncracies are to count in the judgment.

In probabilistic thought, clinical judgment comes full circle, returning to the immediate phenomena, the atomic facts that in themselves have no more meaning than the symptom had, but that, as a totality (like the sum of the symptoms) constitute the disorder. The description of the individual is once again the basis for clinical judgment, except that now the description is an endless series, of which "you have a fever" is barely the beginning.

But the coherence of the data in that series cannot be learned from the print-out that enumerates, even organizes, them. Thus clinical thought, at its seemingly most advanced and comprehensive level, must re-enlist the active collaboration of the individual that characterized its earliest level. In Hegel's sense, clinical judgment has progressed as far as it ever can, that is, it has returned to its origin. It must now, as then, consult the patient for the completion of its judgment. The patient qua individual must enter this cycle of judgments if it is ever to reach a closure, for only the individual can supply the true meanings for the empiricist's collage of facts, the essentialist's categories, and the observer's reading of signs and noting of symptoms.

This does not mean that the dialectic is finally put aside while the patient, alone or assisted, makes autonomous clinical self-judgments. The meaning of 'dialectically complete' is that the outcome or endstage itself has no truth apart from the entire process from which it evolves. Just as medical thought advances through the stages I have described, the participation of the patient must be allowed to progress in the same way. The dialectic of clinical judgment is as incomplete without its early stages as without the later ones. Thus, within each situation in which a judgment is sought as a philosophically complete statement about the individual patient, the person will be assisted to move through the entire progression, together with the clinician (who after all cannot be waiting at the end of the road, since there is no anticipating where any individual's dialectic will lead), both of them regarding each level as necessary to the completion of the whole. This necessity is Hegel's meaning of truth: "the process of its own becoming, the circle which presupposes its end as its purpose;. . . it becomes concrete and actual only by being carried out" ([2], p. 81). "This equal necessity of all of the moments constitutes. . . the life of the whole" ([2], p. 68).

Patient and clinician will therefore need first to fully acknowledge and respond to the immediate symptom as an existential urgency before either of them makes of it a sign. When the sign in turn is made the object of their focus, the patient can learn from the clinician how to use otherwise abstract categories of nosology and particularities of the lesion, as an essential means of overcoming by objectifying the too pressing immediacies of pain and dysfunction. Finally, in the task of creating as detailed and comprehensive a picture of the 'whole person' as possible, it will be the patient who determines which variables and data will be thematic in constituting the individuality that is sought.

It is this process as a whole which is necessary for clinical judgment to be philosophically complete, that is, dialectically developed by clinician and patient and thus directly meaningful for the individual upon whose life it will bear.

Johns Hopkins University, School of Health Services
Baltimore, Maryland.

BIBLIOGRAPHY

1. Foucault, M.: 1975, *The Birth of the Clinic: An Archaeology of Medical Perception*, transl. by A. M. Sheridan Smith, Random House Vintage Books, New York.
2. Hegel, G. W. F.: 1967, *The Phenomenology of Mind*, transl. by J. B. Baillie, Harper & Row Publishers, New York.

THOMAS E. HILL

On first hearing I found the idea of a conference on philosophy and medicine an intriguing one. Metaphors popped readily to mind. Philosophers and physicians: will it be like sports spectaculars where boxers meet Sumo wrestlers with nothing in common but the ring in which they each do their own thing? Or will it be like the even more bizarre TV specials where weight lifters join with marathon runners in a common game of ping pong at which both are incompetent. I refused to entertain the less friendly images: such as a meeting of chemists and alchemists – or even scientists and sages.

Happily, this conference has called attention to several grounds of mutual interest: most notably, questions about the methodology of clinical judgment – the relative merits of statistical analysis, reliance on subjective reports of patients, and so on. I come to the subject from a different perspective, namely, moral philosophy and value theory in general. From this viewpoint, the best I can do is to call attention to another possible area of mutual interest: namely, the place of value judgments, or assumptions, in clinical practice.

Clinical judgment presents problems, as this conference has revealed, that are peculiar to medicine, or anyway to applied sciences. But viewed in the broadest perspective, clinical judgment is just one instance of a larger sort of problem, namely, how to decide what is the best, or most reasonable, thing to do in complex situations. The situations are complicated by the magnitude of relevant scientific data, the uncertainties of outcomes, and the limitations of both computer formulas and individual judgment. But this is not all. To decide what treatment to prescribe a physician must weigh factors of different types: in particular, facts and values.[1] That is, a choice must be made, as in many other areas of life, against a background which includes not only data, hypotheses, predictions and probabilities but also preferences, evaluations, and principles. The scientific complexities, I suspect, are usually the prime focus of attention, perhaps for two reasons. First, they are evident and crucially important; second, they are what one is trained to cope with in a reasonable, reflective way. But to ignore the other aspect, the value assumptions, would be a mistake – in medicine, as in any area of life where decisions have momentous consequences.

H. T. Engelhardt, Jr., S. F. Spicker and B. Towers (eds.), Clinical Judgment: A Critical Appraisal, 254–258.

In medicine, I take it, one needs to size up a patient's condition by asking several questions. When a disorder is suspected, we want to know how extensive it is, what parts it affects, whether it can be cured, and if it can't be cured, how can one best live with it. I want to pose analogous questions about the place of values in clinical judgment: What areas do they affect? Can we get rid of them? And, if not, how can we reasonably live with them?

First, and most obviously, a clinician's values may interfere with his task of assessing the facts of a situation. Maybe dislike of a whiny, fussy old lady may cause one to overlook some real symptom; or sympathy may lead one to underestimate the need for a particularly unpleasant therapy. Pride in one's original surgical technique may cause one to overrate its effectiveness, just as under-confidence might lead to an unduly pessimistic view of the effectiveness of a radical treatment. Racism might encourage one to fix on certain hypotheses: "It's got to be V.D.: I know how those people live." Or sexism might blind one to an hypothesis: "What! That gorgeous young thing! I didn't dream she could have had *that*!"

My examples, of course, are sheer speculation. The particular ways in which values might bias physicians' assessment of facts, and the extent to which this happens are the province of psychologists. The liability here is, of course, a universal one. The more complex decisions are, the more they are made under the pressures of time, and the more they have to do with people rather than medicines and numbers, the more likely that our values affect our judgment of facts. The problem is not unique, or even most pervasive, in medicine. When, just before the English "'Glorious Revolution", the philosopher John Locke judged that his ancestors had made a limited social contract for a certain form of government, he was probably not unaffected by his hope to justify the fall of the Stuarts; and, when (by Plato's account) Socrates judged the soul to be immortal just before drinking the hemlock, doubtless he was not uninfluenced by his wishes.

Is this type of infiltration of values into clinical judgment harmful and to be avoided as far as possible? One's immediate response is "Yes, of course"; and even on reflection I am inclined to say "Yes" – but not "of course". The doubt arises when we consider the following sort of case. Imagine a doctor who has a special regard for a certain patient, generally thought to be incurable, and whose affection causes him to misread the incredible odds against finding a cure. But, so blinded to the facts, he presses on and finds a cure. If his affection had been less, his judgment would have been less clouded; and if his judgment had been less clouded, he would not have found the cure. What this sort of case shows, I think, is that *in some instances* a lucky result can come from the distortion of judgment by values. But one can never know when this will happen. The practical question is: does it pay, as a matter of *policy*, to *try* to resist value-induced mis-readings of the factual situations? No doubt most, maybe even all, of us would agree that this is a more reasonable policy.

The sort of infiltration of values just mentioned is like some diseases: it is harmful, but on the rare occasion it can have good consequences; it is hard, maybe impossible, to eradicate, but at least it can be contained to some degree. But there are other ways values can affect clinical judgment, ways which are completely unavoidable but not to be deplored. They are simply the background against which any practical decision must be made.

These necessary value judgments fall into two categories.

First, and most obviously, there are values by which one assesses the probable consequences of alternative treatments. The most science can tell us is what consequences in fact may result, and, if we are lucky, what the probabilities are. But a decision as to which are *best* for the patient goes beyond science. To be sure, in ordinary cases the underlying values are uncontroversial: other things equal, life is better than death, to lose a limb is worse than to retain it, to live without pain is better than to live in pain. But other things are not always equal. Sometimes these values conflict, and a hard choice needs to be made. Is it better to live longer in an alien hospital environment or not so long at home surrounded by family and friends? What is the relative value of living alertly with pain and living painlessly in a stupor? Is a higher degree of physical health worth the sacrifice of habitual pleasures of a somewhat harmful sort?

The second category of unavoidable value judgments concerns *principles*. Even if one has a confident judgment about what consequences would be *best* for a patient, the question remains "What should be done?" It is tempting to answer, "What is best for the patient, of course!" But are we really prepared, on reflection, to accept such a blanket answer? Suppose, first, that the patient's wishes conflict with your judgment about what is best for him: which should be the guide? Or, again, suppose that the best that can be done for a patient is quite minimal and it would cost his family an enormous financial burden. In some cases perhaps the interests of future patients should be taken into account – as, for example, when one can recommend a double-blind study with an experimental drug the benefits of which are unknown. To pose a further problem: in conditions where the outcome of new treatments is highly uncertain, should the physician simply aim for the best outcome as a rational gambler might, or should he take the conservative stance that it is more important to avoid injury? In short, we may want to modify the pursuit of what is best for the patient by principles of consent, concern for the interests of others, minimal risk taking, and the like.

Opinions about these matters, however implicit and unreflective, are bound to affect judgments about how to treat patients. They cannot be avoided, and, in contrast to bias with regard to *medical* facts, it makes little sense to say that they ought to be constrained – avoided as much as possible. One simply cannot derive conclusions about what *should* be done from premises that refer only to medical facts: some value premises are essential.

What, then, is to be done, given that one must live with unavoidable value judgments. I think of three possibilities:

First, pretend they are not there. That is, focus entirely on the aspects of clinical judgment that are amenable to scientific method and leave the rest to chance. The problem with this head-burying approach is that it allows the crucial value premises to be determined blindly by tradition, habit, and inclination the roots of which we have never examined.

A *second* possible attitude is to revel in the ultimate subjectivity of clinical judgments, using the inevitability of value judgments as an excuse for making decisions by sheer intuition or personal preference. "After all," you might say, "No one can make value free decisions about how to treat patients – so I might as well just make them in the way that feels good to me." This romantic approach has more or less the same result as the scientistic one: the sources and implications of the value commitments that determine crucial decisions are left unexamined. Value choices are openly affirmed – gladly, or with existential *angst* – but we are left with little understanding of why we make these choices, whether they are consistent, what the alternatives are, and what will follow from them.

A third attitude, I think, is best. This is to affirm the necessity of value judgments, admit that they are distinct from scientific judgments, but then expend some time and energy in examining them. Even if value judgments are not capable of strict proof or disproof, they can be more or less understood, more or less consistent, and more or less based on awareness of their implications and alternatives. Something *like* the scientific commitment to consistency, clarity of thought, impartiality and open-mindedness can be maintained in the study of values as well as of facts.

If this sounds like an invitation to do moral philosophy, it is. Socrates said that an unexamined life is not worth living: well, that is perhaps a bit arrogant, but at least we might say that an examined one is more rational, perhaps even more fully human.

University of California at Los Angeles
Los Angeles, California

NOTE

[1] There are, of course, important, and philosophically troublesome, concepts which have both factual and evaluative components: e.g. disease, mental health, normality. But in paradigm cases the distinction is clear enough for present purposes. Compare, for example, 'He will die within ten minutes if we shut off that machine' with 'To prolong his life further is not worth the cost.' It should also be noted that it is a matter of philosophical controversy whether value judgments are a special sort of judgment of fact, but I shall not enter that debate here. In my remarks I assume only that the sort of value judgments in my examples do not reduce to judgments about the medical facts of a patient's situation.

ROUND TABLE DISCUSSION

BERNARD TOWERS

I want to return at this point to some rather simple and basic truths about "judgment at the bedside" (which is what clinical judgment literally means), and then to state what I think is at the root of our current failure to teach our students how to learn to make sound judgments.

Clinical judgment is complicated, and difficult. If we were not aware of that before the Symposium which produced this volume, we surely are aware of it now. The fact ought not to surprise us. When we work with a patient, we are studying the most complex arrangement of matter that nature, so far as we know, has ever produced. We watch this amalgam of organic substances undergoing the strangest of actions and interactions. Pathological processes sometimes simply take over and pursue a relentless course to destruction; but first, and normally, they stimulate a host of autoregulatory defense mechanisms at all levels: simple chemical and physical responses; subcellular activity at the macro-molecular and organelle levels; cellular and tissue reactions; organ-reactions and complicated multi-organ-system responses. Moreover, it is undeniable, in the human species at any rate, that psychic events, both perceptual and affective, often play a vital role in the success or failure of the body's powers of recovery. Simply to understand what is going on is a difficult-enough task for the physician. Further to judge what kind of intervention might help the process of recovery, adds another dimension to our responsibilities. It is even harder to make the judgment, as sometimes we must, that active intervention might impede recovery, and add to our patient's distress: when, in other words, it behooves us to find the courage and the humility to desist, to leave the body to those natural defenses that have developed during the course of evolution, or to leave it to a pathological process that has overwhelmed them.

In the practice of clinical medicine, judgment is involved at every stage: first, in *observation*, including history-taking, where it is crucial first to decide, in other words to judge, what is worth observing or recording and what one can afford to ignore; then in *diagnosis*, about which much (almost too much) has been said during this symposium; next in *therapy*, whether to intervene or not, and if to intervene then in what ways and how often; then repeatedly during the *course of treatment*, when and how to vary the

H. T. Engelhardt, Jr., S. F. Spicker and B. Towers (eds.), Clinical Judgment: A Critical Appraisal, 259–264.

regimen; and lastly (though it often comes first in time and always comes first in importance for the patient) in *prognosis:* what will be, or is likely to be, the outcome? Will I get better, doctor? How long will it take? Shall I modify my plans for next week, next month, next year? The sequence, so far as clinical judgment is concerned, follows a pattern: judgment, observation, judgment, decision; judgment, observation, judgment, decision; then judgment again. And all the time, let me remind you, we are judging behavior-patterns in the most complex arrangement of matter that we know to exist.

A distinguished professor of particle physics once remarked to me that the principle reason for the success of his scientific discipline was that it was basically so simple. He then commented that satisfactory scientific investigation of even the simplest form of living matter must, through lack of ability to control all of the innumerable variables, be virtually impossible. Now we know that modern physicists are currently grappling with very difficult philosophical problems. If that is true of a relatively simple science, it is not surprising that this transdisciplinary symposium between philosophers and physicians has generated a special confusion of its own.

We will not, of course, give up and admit defeat merely on account of difficulties and some confusion. There are two distinct motivations that spur the practitioner of clinical medicine: the first is the fact that when patients get sick, we are often able to help them get better. Second, there is the intellectual appeal of facing up to undeniably difficult problems, and of struggling to resolve them. We succeed sufficiently often to feel a measure of satisfaction both for ourselves and for our patients. When we fail, or when we succeed in spite of, rather than because of, our ministrations, then the appropriate response would seem to be a fitting humility.

Physicians are not noted for humility. Arrogance seems more the mode. Arrogance and cynicism seem to me to be deliberately fostered by the education-system currently in vogue for physicians. Where modesty, and a determination to learn more, might be the appropriate reaction to self-realization concerning one's own ignorance, those virtues often seem to be replaced, these days, by cynicism. Those of us who are medical educators would do well to consider how this comes about in our schooling-system. For one thing we make our students work like galley slaves to master a compendium of facts and formulae which are often trivial and inconsequential. The multitudinous data they are required to commit to memory are regarded by students either as eternal verities or as temporary baggage that they have to carry with them on the tedious journey towards medical licensure. What matter if the 'truths' incorporated into this year's multiple-choice National Medical Board examinations are recognized as false next year or next decade? The standard, deeply cynical reply is that the requirement of continuing medical education will take care of the problem. The units of course-credits, now required at the post-doctoral level, will ensure that all relicensed physicians are up to date in their knowledge of the new and current facts, fads and fancies.

The problem, as I see it, is that the learning-process as currently practised is, by and large, counter-productive for the development of powers of thinking and therefore sound clinical judgment. Increasingly, our teaching and examining in both basic and applied medical science is based on (1) reductionist science, or rather scientism; (2) a discredited philosophy of inductivism; and (3) a simplistic behaviorist psychology that stresses stimulus-response theory as the only effective basis for both teaching and learning. Authoritarian books of multiple-choice review questions are rapidly replacing old-style textbooks, written in continuous prose, as major learning tools for students at all levels of pre and post-doctoral training. The reason is obvious: success in multiple-choice type examinations is rapidly becoming the most potent screening-device of all time for admission to each next professional level. So far as the practice of medicine is concerned a student of today must learn to jump these hurdles, or run these mazes, at every stage from grade-school through college, through medical school, through residency training, and beyond to periodic relicensure for secondary and tertiary specialty-practice.

A multiple-choice test consists, by and large, of a multitude of simple-worded statements of which, on average, about 25% are predetermined to be "true" and about 75% to be "false". The percentage of time and effort expended by examiners in thinking up "true" or "correct" answers, is much less than that spent in thinking up "false" or "incorrect" ones. The lies that have to be invented cannot be just any old lies, or the test will not do its job. They must be subtle lies, sophisticated untruths, designed, as it were, with malice aforethought to trap the unwary or inadequately-conditioned candidate. The exercise is, in my opinion, morally corrupting, and intellectually obscene.[1]

Students tell me that it does not pay to pause and *think* when taking such "objective tests". What you have to do, they say, is to "psych out" the test and, ignoring content and contexts, respond by simple reflex. No credit is given to the candidate for recognizing ambiguity in a question or in the mix of suggested responses. Under these circumstances actual thinking can be positively harmful to academic success, which is absurd. The system requires only that one be well trained to react in stereotyped fashion to preset stimuli.

This makes a mockery of the concept of education in the exercise of judgment. Students ask anxiously, "what am I responsible for?" They want specifics. Out of the total mass of current biomedical data, they want only to be given a selection of those they must remember. To satisfy these pressing needs, teachers are expected and requested simply to provide lists of required data, and not to pose problems of meaning. Once students have run each successive maze, it is not surprising that they often seem glad to be able to dismiss the whole "learning experience" as simply a bad dream or a bad "trip", and glad to tune out and drop out of learning altogether, unless it be by simple exposure to practical experiences.

The word education is derived from Latin. The Romans used two verbs which, in the first person singular, were spelled the same, *e – d – u – c – o*. One has a short e and a long u, the other a long e and a short u. The first is pronounced *ĕdūco* and is conjugated *edūcere, eduxi, eductum*. The other is *ēdŭco*, which conjugates as *ēducāre, ēducāvi, ēducātum*. The first is a typical leadership word, authoritarian, and coercive – to lead out, to train, to discipline whatever it is one has to discipline: troops, a child, employees, a prisoner, a pupil. The second is a caring word, which means to nourish, to sustain, to help to grow. The second word is the true root of our word education. If it were not so we would speak of 'education', as we speak of 'seduction'. Now it is the case that there was some confusion between *ĕdūco* and *ēdŭco* even amongst classical authors, not all of whom were purists in the use of words. Moreover some confusion of meaning is understandable: after all, a measure of discipline is often important in giving good care. The Oxford English Dictionary gives four current meanings for the word 'educate'. They are listed in the order of their evolution, of their historical usage in literature and common speech. The first two, and number 3 in part, are related to caring: examples of their use are quoted from the sixteenth and seventeenth centuries. The later meanings 3 and 4 relate to training and discipline: the first examples of such usage go back no further than the mid-19th century, to the development of Arnold's version of the English Public School System (which means Private in America, of course) with its cold baths, compulsory sports and corporal punishment, all to be endured by the pupil with a stiff upper lip. The system may have trained good officers and public servants for the British Empire. It is not the way, it seems to me, to educate thoughtful and caring physicians.

Organizations such as the Educational Testing Service (ETS) represent, in my opinion, a serious threat to true education.[2] They stultify thinking and reflection, and by their authoritarian dogmatism they stifle the development of powers of judgment. One asks the ancient question that should be asked of all authority: *Sed quis custodiet ipsos custodes*? ("But who shall guard the guards themselves?") Currently, in American education, one can only answer, "No one, except the guards themselves. Or others just like them." This powerful oligarchy (ETS) is self-appointing, self-perpetuating, and describes itself as non-profit-making: the last, I suspect, is true in the strict legal sense only. The phrase is a euphemism, designed, like its location in Princeton, to persuade us of its respectability.[3] The powers of the organization are awesome, and from its rulings there is no appeal. Its system of testing depends, more than any other, on secrecy and confidentiality. If it is the case that there is only one supposedly correct answer to each question, and if "good" (that is, "clever") questions must of necessity be used repeatedly because really "good" ones are in short supply, then there must be safeguards against leaks. But since the relationship that is set up between candidate and examiner is, fundamentally, an adversarial one, and since it is based on deceit and lying, then clever students and clever trainers will discover ways of beating the system. For instance information about the new MCAT (Medical Colleges Admission Test) was published only two months ago. Already there are private cram-schools offering courses of instruction on how to succeed in the test – at tuition costs of between $250-$500. For the aspiring physician this is only the beginning. The MCAT is only the first of many such hurdles. The physician of today must be prepared to be the subject (or object, rather) of an endless series of similar thought-destroying tests in a lifetime of licensure and relicensure.

It is important, of course, to have an accurate data-base, and to have it within relatively easy recall. But it is even more important, if one is concerned to develop judgment-skills, to know how to process the data. Committing facts to memory is necessary, but it is never sufficient. Comprehension is infinitely more worthwhile. It is comprehension, or understanding, that must inform clinical judgment. At this point in the history of education the system we have invented, or have permitted to gain control, largely frustrates and stifles understanding.

In my opinion this is not the way to develop skills in the exercise of clinical judgment. I hope that this symposium, and the publication of its proceedings, will initiate a re-evaluation of what we are or should be about in medical education.

University of California at Los Angeles, School of Medicine
Los Angeles, California

NOTES

[1] For further analysis of the evils of multiple-choice examinations see [1] and [2].

[2] It was only after writing this essay that I became aware of Banesh Hoffmann's book *The Tyranny of Testing* (Crowell-Collier, 1962), based on a series of well-publicized and well-debated articles in *The American Scholar, Harper's Magazine* and *Physics Today*. It is encouraging in one sense to know that there has been other powerful criticism of the system. On the other hand it is depressing to read, in Jacques Barzun's forward to the book, the following assessment of the situation nearly twenty years ago: "Now the tide has turned. As the present book shows, it is the testers who are on the defensive, fighting a rear-guard action against the irresistible force of the argument which says that their questions are in practice often bad and in theory very dangerous". The growth of the system during the intervening years has been remarkable. Scholarly articles and books are clearly inadequate to deal with a problem that is probably economic and political at bottom.

[3] This is not to impute dishonest motives to individual members of the organization, many of whom care deeply about accuracy and scholarship. It is the system itself which is corrupt and corrupting. In fact,the most distressing aspects of this area of modern professionalism are that (1) highly trained and efficient experts seem not to recognize the harm they might be causing, and (2) members of faculty are so easily persuaded to hand over responsibility for judging the abilities of their students to outsiders. The issue is one of ethics and human values, not one of mechanical efficiency.

BIBLIOGRAPHY

1. Towers, B.: 1969, 'Personal View,' *British Medical Journal* **2**, 443.
2. Towers, B.: 1978, 'Medical Education: A Critical Diagnosis', *The New Physician*, **47** (1), 25–26.

H. TRISTRAM ENGELHARDT, JR.

CLOSING REMARKS

This series of symposia has explored conceptual issues in philosophy and medicine. Our hope has been to address the issues that underlie and frame bioethics and medical ethics. We have devoted only one symposium to medical ethics [18], because we have been convinced that the ethical issues raised by medicine can be fully understood only if medicine as science and technology is itself better understood. We were and are convinced that bioethics addresses only a small portion of the important value judgments that structure and guide the health care professions. We have, as a result, forwarded a conceptual analysis of the philosophical issues raised by medicine – what could be termed the philosophy of medicine. Even if the questions which arise in medicine from the philosophy of science, the philosophy of biology, the philosophy of mind, and moral philosophy have no unique articulations with regard to medicine, they should still benefit from being treated together – if not because of some conceptual consanguinity, then at least out of heuristic reasons [7, 9, 17]. We have been attempting to suggest the lineaments of a philosophy of medicine.

The papers in this symposium focus on the pursuit of the philosophy of medicine in the sense that they lay bare some conceptual presuppositions of clinical reasoning. The critical appraisal of clinical judgment which these sessions offer has, I believe, offered a helpful analysis of some of the ambiguities of this term. On the one hand "clinical judgment" is meant to single out judgments concerning medical diagnoses, prognoses, and treatment. On the other hand, "clinical judgment" is, at times, invoked as a refuge in intuition, or poorly explicated reasoning. Both of these sets of issues have been joined in this critical appraisal of the nature of clinical judgment. The focus of these papers has been upon the meaning of knowledge claims in medicine and, to some extent, upon ontological issues (i.e., what there is to know in medicine). These are ancient philosophical concerns of medicine. Interest in the classification of diseases (or nosologies) and the search for alternatives to such classifications reach at least to the disputes between the school of Cnidos (which wished to term each variation in clinical findings a disease entity), and the school of Cos (which engaged in less classification and was more concerned with the variables determining pathological phenomena) [13].

H. T. Engelhardt, Jr., S. F. Spicker and B. Towers (eds.), Clinical Judgment: A Critical Appraisal, 265–271.

These differences were developed in the disputes between ontological theorists of disease (who held that nosologies describe real species of disease or that diseases are discrete things), and the physiological or functionalist theorists of disease (who held that nosologies are classifications of morbid phenomena reflecting the interests of patients, not real divisions of nature) [7]. The first position implied that diseases are real entities. The second position implied that only individual patients are real, and that nosologies primarily reflect human goals and purposes. Modern arguments in the first genre are found in Thomas Sydenham's (1624-1689) *Observationes Medicae* (1676) [19] and in the works of those who followed his lead in holding that clinical findings attest to diseases existing in natural kinds. The *Nosologia Methodica* (1768) [16] of Francois Boissier de Sauvages (1706-1767) and Cullen's *Synopsis Nosologiae Methodicae* (1769) [6] are such works. Diseases, so they argued, exist as perduring types that are merely displayed in different patients. As Sydenham put it " . . . Nature in the production of disease, is uniform and consistent; so much so, that for the same disease in different persons the symptoms are for the most part the same" [20]. Others, including Broussais [5] and Wunderlich [23], criticized such rigid classifications. In fact, many physiological or functionalist theorists of disease were of the opinion that variations in medical phenomena simply reflect different substitutions for physiological variables.

This dispute between the ontological and physiological theorists of disease involved a disagreement concerning the significance of the predictions and control offered by medicine: theirs was a disagreement about the nature of the explanations medicine offers. The functionalists or physiologists were contending that medicine could be reduced to physiology (or at least to a physiology amplified by reference to environmental circumstances). The ontologists held that law-like explanations in medicine cannot be reduced to physiological or other such laws. Disease entities, particular constellations of morbid findings, have, so they argued, a reality which is describable only in the special terms of medicine (or at least such is the position implicit in many of their writings). In short, there was a fundamental divergence in the understanding of the significance of classification, its alternatives, and the bases available for the prediction of outcomes and the provision of therapy.

One must take note, as well, of the difference between clinical accounts and clinical-pathological accounts. That is, one must distinguish among: 1) classifications based on clinical findings, such as those advanced by Sydenham and Sauvages, which grouped illnesses under such basic rubrics as fevers, fluxes, cachexias, weaknesses, etc.; 2) classifications based on anatomical-pathological changes, interpreted as the origin or seats of clinical findings, as suggested by persons such as Xavier Bichat (1771-1802) [2] and Rudolph Virchow (1821-1902); 3) classifications based on etiological agents and mechanisms. In virtue of the second step one ceases to talk simply of jaundice as a clinical entity and begins to speak of jaundice in terms of particular bases (e.g., hepatitis). In doing so one moves from a purely clinical characterization of disorders to a characterization in terms of underlying pathological findings. In virtue of the third step, instead of simply speaking of hepatitis, one speaks of viral hepatitis, toxic hepatitis, immune hepatitis, etc. Also, clinically diverse phenomena such as consumption and Pott's disease become grouped as manifestations of tuberculosis because of an accepted explanatory model.

Clinical judgment is, therefore, more for us than an eidetic phenomenology of syndromes in the mode of Sydenham, Sauvages and Cullen. Modern medicine has provided explanatory accounts which support and restructure the significance of our clinical classifications [11]. We are confronted, then, with the problem of maintaining our concern with the presented character of diseases following Sydenham, while providing classifications of disease which reflect our explanatory schemes framed in terms of pathological and etiological considerations [21]. Clinical judgment, as Feinstein has argued, requires that we have classifications in both clinical and pathological terms. [10].

As these developments indicate, the character of clinical judgment has changed in a dramatic fashion over the more than three hundred years since the publication of Sydenham's *Observationes Medicae* (1676) [19], so that clinical findings are now reinterpreted in terms of the pathological-anatomical accounts of such phenomena. One must remember that one of the reasons Sydenham and others engaged in clinical classifications was to provide an empirical medicine free of a priori speculation. In addition, more sophisticated recipes are now available for decision-making in clinical contexts. Our interest in such decisions is not however *de novo*. The recognition of the need for greater rigor in clinical reasoning has roots that go back, if not to Hippocrates [12], then at least to the medical logics of the nineteenth century such as Blane's *Elements of Medical Logic* [4], Bieganski's *Medizinische Logik* [3], and Oesterlen's *Medizinische Logik* [14].

Much, though, has changed. Medicine is now a fairly well developed science (in the sense of possessing considerable explanatory power) and a fairly successful technology (in the sense of enabling impressive control of morbid states). Up until recently, however, medicine succeeded more in being descriptive than explanatory, and thus in being a set of techniques rather than a technology (in the sense of successfully applying a useful, general scientific account, where success is measured in terms of favorably influencing the health of patients and usefulness is measured in terms of helping guide techniques to success while accounting for why such techniques are successful). In this volume we have taken some first steps to distinguish, at least implicitly, between medicine as a science and medicine as a technology, and to explore each of these dimensions. It is one thing to understand how medicine frames classificatory systems and another to understand how one employs recipes for categorizing individual patients and attending to their cure. As Professor Ernan McMullin indicated in unpublished remarks at the symposium, the reasoning of physicians is usually in one sense closer to that of engineers than to that of scientists (a comparison meant to be neither derogatory nor laudatory). Unlike the scientist who wishes to understand the general laws of physics, the properties of materials, and of seismology, etc., for their own sake, the engineer is concerned with such things as the stability and trustworthiness of particular bridges. General knowledge is appealed to for its use in particular cases. Analogously, unlike the biomedical scientist who may be concerned with the general properties of different infectious agents in humans, the clinician is concerned with the properties of particular pathogens in particular human patients.

These distinctions may be drawn in different ways. One can identify a clinical *science* of medicine which attempts to understand the ways in which classifications of morbid phenomena are made, and a clinical *art* which applies to individual patients those classifications ard rules for identifying morbid phenomena. Clinical medicine in the first sense is like the science of engineering; clinical medicine in the second sense is like the art of practicing engineering. Thus, there is sense in saying that each relies upon a type of clinical judgment – the first upon decisions as to what will in general be useful in the clinical context, the second upon decisions as to what will be useful with respect to a particular patient. We may then distinguish a clinical science of medicine, an art of medicine, and a rigorous account of each of these two enterprises. That is, one can distinguish: 1) the general study of the morbid phenomena displayed by patients in clinical settings; 2) the application of such knowledge by physicians to particular patients, which application can be termed the art of medicine; 3) a rigorous account of the clinical science of medicine (i.e., a philosophy of the clinical science of medicine); and 4) a rigorous study of the criteria invoked by clinicians in practicing their art (i.e., a philosophy of the art of medicine). For the most part, this symposium has focused on the fourth enterprise, the rigorous study of the art of medicine. The goal has been to provide a better understanding of that art. In examining classifications and their alternatives, the lineaments of an examination of the science of clinical judgment or clinical practice has been offered. The art of medicine itself has also been examined. But the distinctions between these two critical appraisals have only been imperfectly suggested. It will be necessary to draw them out in greater detail in future symposia.

Finally, medicine is a science and technology with profound direct and indirect consequences for persons. The goals of the sciences and technologies of medicine are more intimately embedded in the matrix of human values than, say, the goals of the science of astrophysics, or the technology involved in retrieving moon rocks. Medicine begins in and returns to human complaints. Considerations of the sciences and technologies of medicine are, thus, intimately bound to issues in bioethics. In fact, there are evaluations which are prior to the evaluative judgments that constitute bioethics. The characterization of conditions as pathological turns on what humans perceive to be states of suffering. To call a state of affairs a disease or illness involves identifying that state as abnormal or in some sense improper – it involves a value judgment [8]. Consider, for example, how polydactylia is held to be an abnormal or a pathological state even when unassociated with other findings. Consider also the nature of the decision regarding whether menopause should be considered a disease or a natural state of affairs [1]. Such decisions are not simply descriptive. As the discussions in the first volume of this series indicate, such decisions are dependent upon evaluations ([15], [22]).

Where does this leave us in our consideration of the nature of clinical judgment? I believe we are brought a step further in the understanding of the sciences and technologies of medicine, an effort in understanding with a considerable history. We must still undertake further analysis of the sciences and technologies of medicine, and their bearing upon and inspiration from human goals and values. We are led also to the need for further understanding the distinction among the science of clinical medicine, the art of clinical medicine, the rigorous study of the science of clinical medicine, and the rigorous study of the art of clinical medicine. The nature of these various enterprises and the character of their differences will be the subject of future symposia in this series.

Kennedy Institute of Ethics
Georgetown University
Washington, D.C.

BIBLIOGRAPHY

1. Barnes, A. P.: June, 1962, 'Is Menopause a Disease?', *Consultant* **2**, 22–24.
2. Bichat, X.: 1801, *Anatomie géněral appliquée a la physiologie et à la médicine*, 4 vols., Chez Brosson, Gabon et cie., Paris.
3. Bieganski, W.: 1894, *Medizinische Logik; Kritik der ärzlichen Erkenntnis*, Kowalewski, Warsaw, (transl. by A. Fabin: 1909, C. Kakitzsch, Wurzburg).
4. Blane, G.: 1822, *Elements of Medical Logic*, Huntington and Hopkins, Hartford, Connecticut.
5. Broussais, F. J. V.: 1821, 'Examen des doctrines médicales et des systèmes de nosologie', Mequignon-Marvis, Paris, Vol. 2, p. 646.
6. Cullen, W.: 1769, *Synopsis Nosologiae Methodicae*, Edinburgh.
7. Engelhardt, H. T., Jr.: 1975, 'The Concept of Health and Disease', in H. T. Engelhardt, Jr. and S. F. Spicker (eds.), *Evaluation and Explanation in the Biomedical Sciences*, Reidel, Dordrecht, Holland, pp. 125–141.
8. Engelhardt, H. T., Jr.: 1976, 'Human Well-Being and Medicine: Some Basic Value-Judgments in the Biomedical Sciences', in H. T. Engelhardt, Jr. and D. Callahan (eds.), *Science, Ethics and Medicine*, Hastings Center, Hastings-on-Hudson, New York, pp. 120–139.
9. Engelhardt, H. T., Jr., 1976, 'Ideology and Etiology', *Journal of Medicine and Philosophy* **1**, 256–258.
10. Feinstein, A. R.: 1967, *Clinical Judgment*, Krieger Publishing Company, Huntington, New York.
11. Foucault, M.: 1973, *The Birth of the Clinic: An Archaeology of Medical Perception*, (transl. by A. M. S. Smith), Pantheon Books, New York.
12. Hippocrates, *Precepts* I–III.
13. Jones, W. H. S.: 1962, 'Introduction', *Hippocrates*, Harvard University Press, Cambridge, Mass., Vol. I, p. xii.
14. Oesterlen, F.: 1852, *Medizinishe Logik*, H. Laupp, Tübingen. 1855, *Medical Logic*, ed. and transl. by G. Whitley, C, Kabitzsch, Wurzburg.
15. Pellegrino, E. D.: 1975, 'Round Table Discussion', in H. T. Engelhardt, Jr. and S. F. Spicker (eds.), *Evaluation and Explanation in the Biomedical Sciences*, Reidel, Dordrecht, Holland, pp. 228–234.
16. Sauvages, F. B. de: 1768, *Nosologia Methodica Sistens Morborum Classes Juxta Sydenhami mentem et Botanicorum ordinem*, Fratres de Tournes, Amsterdam.
17. Shaffer, J.: 1975, 'Round Table Discussion', in H. T. Engelhardt, Jr. and S. F. Spicker (eds.), *Evaluation and Explanation in the Biomedical Sciences*, Reidel, Dordrecht, Holland, pp. 215–219.
18. Spicker, S. F. and Engelhardt, H. T., Jr. (eds.): 1977, *Philosophical Medical Ethics: Its Nature and Significance*, Reidel, Dordrecht, Holland.
19. Sydenham, T.: 1676, *Observations Medicae circa Morborum Acutorum Historiam et Curationem*, G. Kettilby, London.
20. Sydenham, T.: 1848, 'Preface', *The Works of Thomas Sydenham*, ed. 3, (transl. by Greenhill), The Sydenham Society, London, vol. I, p. 15.
21. Wartofsky, M. W.: 1975, 'Organs, Organisms, and Disease' in H. T. Engelhardt, Jr. and S. F. Spicker (eds.), *Evaluation and Explanation of the Biomedical Sciences*, Reidel, Dordrecht, Holland, pp. 67–83.

22. Wartofsky, M. W.: 1975, 'Round Table Discussion' in H. T. Engelhardt, Jr. and S. F. Spicker (eds.), *Evaluation and Explanation in the Biomedical Sciences*, Reidel, Dordrecht, Holland, pp. 222–228.
23. Wunderlich, C. A.: 1842, 'Einleitung', *Archiv für physiologische Heilkunde* I, ix.

NOTES ON CONTRIBUTORS

Morton Beckner, Ph.D., is Professor of Philosophy, Pomona College, Claremont, California.

Eric Cassell, M.D., is Clinical Professor of Public Health, Cornell University Medical College, New York, New York.

Arthur Elstein, Ph.D., is Professor and Director, Office of Medical Education Research and Development, Michigan State University, East Lansing, Michigan.

H. Tristram Engelhardt, Jr., Ph.D., M.D., was at the time of the conference, Associate Professor, Institute for the Medical Humanities and the Department of Preventive Medicine and Community Health, The University of Texas Medical Branch, Galveston, Texas. He is currently Rosemary Kennedy Professor of the Philosophy of Medicine, Kennedy Institute of Ethics, Georgetown Univeristy, Washington, D.C.

Sally Gadow, R.N., Ph.D., is Assistant Professor of Philosophy, School of Health Services, Johns Hopkins University, Baltimore, Maryland.

Linda L. Gard, B.A., is Staff Assistant, Department of Radiology, Michigan State University, East Lansing, Michigan.

John L. Gedye, M.B., B.Chir., was at the time of the conference, Research Associate Professor of Neurology, Baylor College of Medicine, Houston, Texas. He is currently Professor of Neurology, School of Medicine; and Adjunct Professor of Electrical and Computer Engineering, College of Engineering, Wayne State University, Detroit, Michigan.

Marjorie Grene, Ph.D., is Emeritus Professor of Philosophy, University of California at Davis, Davis, California.

Thomas E. Hill, Ph.D., is Associate Professor of Philosophy, University of California at Los Angeles, Los Angeles, California.

Martin Lean, Ph.D., is Professor and Director, School of Philosophy, University of Southern California, Los Angeles, California.

Ernan McMullin, B.D., Ph.D., is Professor of Philosophy, University of Notre Dame, Notre Dame, Indiana.

Sherman M. Mellinkoff, M.D., is Dean of The UCLA School of Medicine, Los Angeles, California.

Edmond A. Murphy, M.D., Sc.D., is Professor of Medicine and Director,

Division of Medical Genetics, The Johns Hopkins University School of Medicine, Baltimore, Maryland.

Edmund D. Pellegrino, M.D., was at the time of the conference Professor of Medicine and President, Yale/New Haven Medical Center, New Haven, Connecticut. He is currently President, The Catholic University of America, Washington, D.C. and Professor of Philosophy and Biology. He is also Clinical Professor of Medicine, Georgetown University, Washington, D.C.

E. James Potchen, M.D., is Professor and Chairman, Department of Radiology, Professor of Health Systems Management, Michigan State University, East Lansing, Michigan.

D. L. Rosenhan, Ph.D., is Professor of Psychology and of Law, Stanford University, Stanford, California.

William R. Schonbein, M.A., is Assistant Dean, College of Human Medicine, Michigan State University, East Lansing, Michigan.

Michael Scriven, Ph.D., was at the time of the conference Professor of Philosophy, University of California, Berkeley, California, and is now University Professor and Director of the Evaluation Institute at the University of San Francisco, San Francisco, California.

Elliott Sober, Ph.D., is Assistant Professor of Philosophy, University of Wisconsin, Madison, Wisconsin.

Stuart F. Spicker, Ph.D., is Professor of Philosophy, University of Connecticut Health Center, Farmington, Connecticut.

Patrick Suppes, Ph.D., is Lucie Stern Professor of Philosophy, and Director of Institute for Mathematical Studies in the Social Sciences, Stanford University, Stanford, California.

Bernard Towers, M.B., Ch.B., is Professor of Pediatrics and Anatomy, University of California at Los Angeles, Los Angeles, California.

Paul Wahby, B.S., is Research Associate, Department of Radiology, Michigan State University.

Daniel I. Wikler, Ph.D., is Assistant Professor and Kennedy Scholar in the Program in Medical Ethics, School of Medicine and in the Department of Philosophy, University of Wisconsin, Madison, Wisconsin.

INDEX

The Philosophy and Medicine Book Series

Editors

Stuart F. Spicker

and

H. Tristram Engelhardt, Jr.

1. **Evaluation and Explanation in the Biomedical Sciences**
   Proceedings of the First Trans-Disciplinary Symposium on Philosophy and Medicine, Galveston, Texas, May 9–11, 1974
   1975, vi + 240 pp. ISBN 90-277-0553-4

2. **Philosophical Dimensions of the Neuro-Medical Sciences**
   Proceedings of the Second Trans-Disciplinary Symposium on Philosophy and Medicine, Farmington, Conn., May 15–17, 1975
   1976, vi + 274 pp. ISBN 90-277-0672-7

3. **Philosophical Medical Ethics: Its Nature and Significance**
   Proceedings of the Third Trans-Disciplinary Symposium on Philosophy and Medicine, Farmington, Conn., Dec. 11–13, 1975
   1977, vi + 252 pp. ISBN 90-277-0772-3

4. **Mental Health: Philosophical Perspectives**
   Proceedings of the Fourth Trans-Disciplinary Symposium on Philosophy and Medicine, Galveston, Texas, May 16–18, 1976
   1978, xxii + 302 pp. ISBN 90-277-0828-2

7. **Organism, Medicine, and Metaphysics**
   Essays in Honor of Hans Jonas on his 75th Birthday, May 10, 1978
   1978, xxvii + 330 pp. ISBN 90-277-0823-1